The Pre-scientific Bible

A study of cultural influences on the biblical writers, and how these affect our reading of the Bible today

The Pre-scientific Bible

A study of cultural influences on the biblical writers, and how these affect our reading of the Bible today

Peter Colyer

Circle Books

Winchester, UK
Washington, USA

First published by Circle Books, 2013
Circle Books is an imprint of John Hunt Publishing Ltd., Laurel House, Station Approach,
Alresford, Hants, SO24 9JH, UK
office1@jhpbooks.net
www.johnhuntpublishing.com
www.circle-books.com

For distributor details and how to order please visit the 'Ordering' section on our website.

ISBN: 978 1 78099 914 2

A CIP catalogue record for this book is available from the British Library.

Design: Stuart Davies

Printed and bound by CPI Group (UK) Ltd, Croydon, CR0 4YY

We operate a distinctive and ethical publishing philosophy in all areas of our business, from our global network of authors to production and worldwide distribution.

CONTENTS

To all those who love the Bible
but do not read it thoroughly enough

1

What this book is about

The Bible and Science

The Bible is the source text for the Christian faith. But the world in which the Bible texts were produced and compiled, over a period of several centuries, was enormously different from the present world of the twenty-first century. In particular, the ideas and assumptions about the natural world and its meaning have passed through great cultural changes. One of the most significant of these changes has been the emergence of experimental science as the basis for explanations of what happens in the physical world.

This shift in our cultural mentality raises the question of how sacred texts that arose in very different cultural circumstances can continue to be the basis of an active and lively faith. I state at the outset that my answer to this question will be positive, but difficult questions have to be faced and some traditional attitudes to the Bible may have to be revised.

Interest in the relationship between scientific knowledge and Christian faith has increased enormously during the past few decades. This interest has been expressed in the publication of many books aimed at a wide range of readership, and in the creation of new university courses. Among the authors particularly prominent in this development are Ian Barbour in the United States and John Polkinghorne and the late Arthur Peacocke in Britain. Many others have also made worthy contributions to this debate.

However, this growth of interest in the subject has focused almost entirely on what may be described as the philosophical or historical aspects of the relationship. The questions generally addressed may be summarised, with some simplification, as of the type "Is it reasonable for a scientist to adhere to the Christian faith?" or "Is scientific methodology consistent with religious belief?" or "How has the relationship between faith and science been understood in the past?" These questions and the answers offered are of great value in a society which prizes scientific knowledge and expects evidence to be produced if claims about knowledge or truth are to be accepted.

The nature of much of the popular literature in the recent Science and Religion debate can be illustrated from the promotional words on the back cover of Francis Collins' helpful book *The Language of God*. Collins "takes the reader on a stunning tour of modern science to show that physics, chemistry and biology – indeed, reason itself – are not incompatible with belief."[1] The conclusion "not incompatible" is typical – the discussion is not about absolute proofs but about probabilities and alternative means of arriving at working conclusions.

During the twentieth century science moved away from the confidence of the Newtonian world to the more difficult world of Einsteinian relativity, Heisenberg uncertainties and quantum mechanics. In many ways this transformation of the scientific approach opened new opportunities for dialogue between scientific ideas and religious beliefs. The relationship between these two forms of knowledge was no longer as clear as the more mechanical nineteenth century science had assumed. The analysis of these probabilistic trends in scientific thinking has been another valuable theme in recent literature. However, in this book I do not intend to examine these scientific issues in detail – they are part of the assumed background. This book is primarily about the Bible and its place as a foundational religious text in a scientifically dominated society.

The place of the Bible in the relationship with science has been a neglected theme in these recent developments. Christian faith has generally been treated as a "known", so questions about its relationship with science can be framed in the forms described above. However, the Bible itself is a foundational element in the Christian faith, and *the increased knowledge provided by the scientific endeavour also raises questions about these source documents*. It should not be assumed that there is a fixed phenomenon called Christianity, which remains unchanged, while all that requires investigation is the relationship of Christian faith to modern science. The Bible itself has to be examined in relation to scientific thought – this exercise may reveal not only that the relationship between Christian faith and scientific knowledge is changing (or should change), but also that Christian faith is itself changing because of the impact of scientific thought on the way Christians understand the Bible.

The books of the Bible were written over a period of many centuries, between several hundred years BCE (though some individual texts may be considerably older) and the first century of the Christian era. Human beliefs and worldviews during all this time may be described as "pre-scientific". The emergence of science, not only as an investigative activity but also as a mode of rational thought based on empirical evidence, is usually regarded, at least in the Western world, as beginning in the sixteenth and seventeenth centuries.[2] A crucial role in bringing about this mental change was played by the recognition that the planetary system circulated around the Sun and not the Earth. The discovery of the heliocentric nature of the solar system was particularly associated with the Polish mathematician-astronomer Nikolai Copernicus (1473-1543) and the Italian mathematician-engineer-astronomer Galileo Galilei (1564-1642). In numerous ways this new perception moved European thought away from the medieval view of the world inherited from classical times.

Expanded knowledge of the world's geography and topography also arose from the voyages of maritime explorers from the late fifteenth century onwards – the explorers brought home examples of plants and animals, and information about other peoples and cultures, that had no place in the prevailing European view of the world, which was derived from biblical or classical sources. This new knowledge contributed along with the findings of the astronomers to the profound revision of long-held ideas. Among the mental changes gradually introduced into Europe by these discoveries were the recognition that the Earth was not the physical centre of the universe (with consequent questions about the place of the Earth in God's purposes), that heaven and hell could no longer be regarded as physical locations at the extremities of height and depth respectively, and that there was much more variety in the world and its peoples than had previously been recognised.

To describe the pre-sixteenth century world as "pre-scientific" is not intended to denigrate the knowledge and many practical skills exercised in earlier societies. Greeks, Romans, Egyptians and other cultures from the Middle East and further east had achieved striking advances in philosophy, poetry, drama, mathematics, astronomy and the practical skills required for the construction of buildings, navigation across open seas, and the government of large extents of territory. Some of these discoveries were foundational for the whole of subsequent Western knowledge. Describing these societies as pre-scientific is simply a recognition that before the emergence of empirical science, the physical causes of events were largely unknown. As Keith Ward has written:

> ... until the seventeenth century even the most intelligent observers were not generally aware of the true causes of most of the physical changes in the universe. That is to say, virtually no one thought that changes occurred in accordance with universal, mathematically quantifiable laws of nature ... for most of human history even those who most assiduously sought to discover the true causes of change were mistaken about those causes.[3]

These ancient cultures had developed to the point at which individuals could be released from the basic requirements of subsistence living to concentrate their productive efforts on specialised activities for which they were particularly skilled – such a specialisation of employment is an essential ingredient for urban living and the development of "civilisation". These specialisations might include, for example, metalwork or soldiering or building, enabling others to concentrate on leatherwork or the manufacture of clothing or the processing of food. Activities beyond the basic needs of living will also emerge from this process of specialisation. Thus in more developed societies there will be some people who will be enabled to concentrate on activities such as the writing of poetry or the creation of music.

Some ancient societies were well endowed with technological skills in fields such as agriculture, natural medicines, building and construction, the manufacture of tools and weapons, and in some cases seafaring. The societies described in the one or two thousand years of biblical history possessed many of these capabilities. But these substantial achievements were limited in two important respects: first by the size of what could be seen by the human eye, and second by the weight of what could be handled by human or animal muscle power, assisted to some extent by ropes, pulleys and levers: ancient technologies were limited by what could be seen and handled. The inner workings of things were not well understood. For example, the functions of the human body and the nature of disease were known only in the broadest outline: it was not until the seventeenth century CE that the circulation of the human blood system was clarified by William Harvey (1578-1657). The influences which lay behind the patterns of the weather and the consequent variations in agricultural output were not known until, also in the seventeenth century, the fundamental importance of atmospheric pressure had been established, initially by Evangelista Torricelli (1608-

1647), Galileo's assistant and successor in Florence. In earlier centuries very few measuring instruments were available, partly because of the difficulty of manufacturing to the required precision but more importantly because of the absence of awareness of the questions which needed to be asked and the values which needed to be measured.

When we consider the geographical location of the historical and religious events described in the Bible, it is remarkable that ancient Israel and the Jewish people remained in a relatively undeveloped state while surrounded by a series of sophisticated civilisations. These surrounding nations had enormous successive impacts on the story of the Jewish people, which makes it even more surprising that Israel seems to have absorbed little of the technological achievements of her neighbours. At one stage the early forebears of the nation, or at least a sub-group within the alliance which later became the nation of Israel-Judah, were slaves in Egypt. The achievements of Egypt in constructing massive temples, palaces and pyramids, and maintaining a complex society based on intensive agriculture, are well known, and according to the biblical account the Israelite slaves participated in some of this construction work. Yet when they escaped and established a nation and a land of their own, centuries elapsed before Israel could boast any equivalent buildings, and even then very few. The kingdoms of Israel and Judah were then successively overwhelmed by the powerful nations of Assyria, Babylon and Persia – the Babylonians made substantial progress in the fields of astronomy and mathematics, but little interest in these fields is found in our knowledge of the Hebrew people. Later still, in the period between the completion of most of the Hebrew Scriptures (the Old Testament) and the events described in the New Testament, the Jewish nation was conquered by the empires of Greece and Rome, with their achievements in philosophy and literature, architecture, military power and social organisation. The Greek philosopher Aristotle (384-322 BCE), who of all the classical Greek writers approximates most closely to the later pattern of an investigator of the natural world (though Aristotle's method was primarily observational and philosophical rather than experimental), lived during the period approximately represented by the gap between the Hebrew Scriptures and the New Testament.

Yet, in spite of these associations with some of the most technologically advanced nations of their times, the nation of Israel appears to have absorbed little of the technological drive of their conquering neighbours, at least as far as we can detect from archaeological exploration and from their literature preserved in the Bible. The main contributions of the ancient Israelite people were in the realm of ethics and religion. Perhaps the association of relatively advanced technology with nations frequently regarded as enemies served to insulate Israel from moving in this direction. We must also recognise that the literature preserved in the Bible may be an unrepresentative cross-section of what was happening at the time, first because the Hebrew Scriptures deal mainly (though, as we shall see, not entirely) with religious topics which may exclude information about the place of technology, except insofar as this is revealed by reading between the lines; and second because the authors of the texts of the Hebrew Scriptures were, again as far as we can tell, mainly prophets, priests and officials who were not concerned to tell us anything about the nation's interest in technology. In spite of these qualifications, the impression remains that Israel was a technologically backward society surrounded by more advanced empires.

The purpose of this book

The purpose of this book, therefore, is to examine how scientific knowledge and scientific methods of thought relate to our understanding of the Bible. This exercise should undergird the wider debate about the relationship between science and the

Christian faith.

Since my purpose is to examine the biblical texts themselves in comparison with later scientific knowledge, the fact that the texts may not be fully representative of their contemporary society is not a serious disadvantage. The biblical texts, which followers of both the Jewish and the Christian faiths regard as in some way revelatory of God, are the important source material with which we are concerned. The extent to which these texts reveal facets of contemporary life in ancient Israel is an interesting question, but my present purpose is to examine the content of the texts as they stand. The degree to which the texts reflect accurately or fully the contemporary way of life is a separate issue which I will touch on only when it bears on my primary question. In other words, there may have been more technology in the biblical societies than the texts reveal, but the primary interest of Christians is in the texts – the societies behind the texts are of interest in illuminating the texts, which are themselves the source material for Christian faith.

In subsequent chapters we will conduct detailed examination of many texts in both the Hebrew Scriptures and the New Testament, with specific reference to their assumptions about the working of the natural world. My orientation is not solely historical or textual; rather, I am hoping to help present-day believers in their interpretation of our source texts. These texts were created in societies with assumptions very different from our own about the operations of the physical world around us.

A major result of our study of the Scriptural texts will be that the biblical writers believed in *the direct activity of God in the natural world*. The events in question might be weather conditions, earthquakes, conceptions, deaths, and many other occurrences. We now know much more about the physical reasons for these phenomena, and a modern reader wishing to retain a role for God might prefer to render the statement more subtle by saying that God "was responsible for" or "was behind" the event rather than "performed" the event. This difference of outlook can create an intense hermeneutical problem: do we have to accept the worldviews of the original authors and understand the events in the physical world in the same way as they did? Or should we reinterpret the text in conformity with our present scientific understanding of the world? Or is some middle ground possible? If some reinterpretation is necessary, how much of the original sense is lost? Is it possible for a scientifically oriented reader in the twenty-first century to appreciate the sense and impact of the original, while also retaining scientific integrity? These are the crucial questions raised in this book and will be thoroughly explored in Chapter 7 when we have examined all the relevant biblical data.

At this stage a single example will help to illustrate the issue which will be faced many times throughout this book. The Bible text may say quite bluntly that God performed a particular act, while we may prefer to understand that the event happened as a result of a process which had been in continuous operation for some time, perhaps a very long time, before the event happened. The first two chapters of the book of Joel describe in graphic terms a devastating invasion by locusts, and Joel 2.11 states that this is the Lord's army and that the Lord is giving the orders:

The Lord utters his voice at the head of his army;
How vast is his host!
Numberless are those who obey his command.
Truly the day of the Lord is great;
Terrible indeed – who can endure it?
Joel 2.11

In our present times most believers would be reluctant to attribute a locust plague, or any similar form of disaster, to such a direct intention of God. We would prefer an explanation based on climatic factors, recent weather and wind directions, or the demographic pressures on insect populations. Lessons would be learned, and steps taken to mitigate such events in future if possible. How should we resolve this potential conflict of understanding as we read biblical accounts of divine action? This example illustrates the problem of the exercise of Christian faith in a cultural environment very different from that in which the faith originated. Should a Christian accept the bald statement that God did it, or is a more nuanced interpretation possible – and if the latter, is faith diminished or increased? These issues lie behind all the subsequent studies in this book.

Four preliminary issues

Before we progress into detailed themes and texts, four initial issues need to be clarified.

First, I wish to avoid the view that there is a single "biblical" approach to some of the questions we shall address. The biblical texts were created by many different authors over a period of more than a thousand years, say between 1000 BCE and 100 CE, though some parts may lie outside these limits. During such a long period social conditions, religious beliefs and knowledge of the world all changed substantially, and the meanings of words and even the languages used were different (the Hebrew, Aramaic and Greek languages are all employed in the biblical texts). It should not be expected that every author had the same perspective on human experience, on the being of God, or on the relationship between the two. Of course there are uniting themes throughout the Bible, but large differences should also be expected, particularly between the Hebrew Scriptures and the New Testament, but also within these two major parts of the Bible.

Some Christians may feel that the Bible must contain a certain unity because it deals with "the divine plan of salvation" or because it is "divinely inspired" or because it has to be "Christ-centred". These ideas may be correct to some extent, but such considerations should not obstruct a careful analysis of particular texts. The concept of revelation through the Scriptures has been to a large extent superimposed on the texts after they were collected into a single volume. In this book I am hoping to concentrate on the meaning of the texts in the milieu in which they first arose before addressing their interpretation in the light of scientific knowledge. For the most part, my approach will be to allow the different biblical books, or even sections within books, to speak for themselves without attempting forced harmonisations. When I use the word "biblical" I am therefore intending a simple description of a particular group of texts: I am not implying a single theological viewpoint throughout.

Second, I hope to avoid any impression of using "proof-texts". A glance through the pages of this book will reveal places where I have assembled numerous biblical texts relating to the issue under discussion. In most cases I have quoted the actual texts rather than merely given the verse reference – in this way I hope the reader will be convinced of the strength and frequency of the points I wish to make. This reproduction of texts may give the impression of a list of proof-texts, taken out of their original context within a biblical book and a particular historical setting. I cannot express too strongly my principled opposition to such a practice – my assembled texts are intended as illustrations of a viewpoint held by the writer or writers, not as "proofs" of an idea. I have examined the context of every Scripture quotation used, and have taken account of the latest scholarly views of the background and original meaning. I have not felt it necessary to give an extended exposition of every text – this would not be relevant to my immediate purpose – but I have endeavoured to use every text in the sense intended in its original setting, insofar as that can be established. In some cases this has required a more extended examination of the original sense – this should not necessarily be read as an endorsement in our modern, scientifically oriented situation. In books such as the Psalms and

Proverbs, which contain material for worship in many settings, or wisdom of general application, the exact context will often be harder to detect. In these books it is more likely that the ideas contained were regarded as universal truths, though the cultural background or setting of the worship remains important.

Third, this book concentrates on the presence or absence of scientific knowledge in biblical times and the relationship between ancient and more recent world views. This means that many grander themes in Christian theology such as the nature of "salvation" or the Christian view of atonement will not be directly addressed, though there may be repercussions between the topics dealt with here and larger theological questions. Indeed, it would be a pity if that did not happen.

The *fourth* preliminary question relates to the words "nature" or "natural" or the expression "the natural world". The use of these words may suggest the existence of an independent entity, the natural world, which should be considered separately from the being of God, thus prejudging the question of the relationship of the world to God. The very use of a concept such as "natural" may be already endowed with a hidden assumption of a distance between the divine and the material. Many of the biblical writers themselves may not have recognised the suggestion of a "natural" world other than the world controlled and determined by God. These are issues to be explored in later chapters of this book. Meanwhile, it would be heavy-handed and overbearing to exclude altogether the use of words such as natural, or to introduce clumsy qualifiers such as "so-called natural". For the present let it suffice to say that my use of words and expressions involving the term "natural" means the physical world, including both the non-human world and the physico-biological aspects of the human world. The recognition of a zone described as natural is based on the existence of physical forces immediately responsible for the effects we see. It does not imply a mystical or quasi-personal existence such as "Mother Nature". Humans have to be regarded as natural, at least in our physico-biological existence. I do not wish to prejudge the question of whether there is a world of physical processes operating independently of God, but equally it would make for a very laboured text if I avoided completely the term "natural".[4]

Human and non-human aspects

In this study I will examine the biblical attitudes towards both the human and non-human aspects of the natural world. The two primary focuses of the whole Bible, it may be said, are God's purposes and human behaviour, and how to bring these two together. The historical books, the laws and the prophecies of the Hebrew Scriptures tell the story of God's dealing with one particular nation and how they should live in the land. The New Testament is almost entirely composed of the life and teaching of Jesus, the emergence and growth of the Christian church, and collections of letters giving advice to churches or individual Christians. The main reflective parts of the Bible, the Psalms and other wisdom literature in the Hebrew Scriptures, where a view of nature for its own sake might be found, are also primarily concerned with human life and the divine-human encounter, though it is here that we will find a particularly rich source of information on a biblical view of the physical world.

Because of this overwhelming concentration of the biblical texts on the divine-human relationship, the views of the biblical writers about the non-human world have to be found embedded within what they have to say about human behaviour in society, or about the worship of God. Nevertheless, a wealth of information is available, even though much of it derives from incidental references.

The structure of the book

Chapter 2 addresses the question of where the biblical writers believed that God was to be found. In Chapter 3 I will examine texts concerning the role of God in the non-human world of landscape, animals and the weather, although as we shall see it is ultimately impossible to distinguish between a non-human world and the world as the home of mankind – for the biblical writers the world had been created as a home for mankind. Our modern idea that the world had existed for billions of years without human life was of course unknown to them. In Chapter 4, therefore, we progress to examine the views of the biblical authors about the divine-human relationship that dominate and underlie what they have to say about the physical world. Chapter 5 considers the relationship between miracles and magic, which appears to have changed in character over the period of biblical history. The discussion of miracles leads naturally to a consideration of God's omnipotence.

Chapters 2 to 5 will conclude with a short comparison of the biblical view with a modern, scientifically informed view of the phenomena under discussion. Discussion of the major theological issues will be retained until Chapter 7.

In Chapter 6 I explore the presence in the texts of both early ideas that might loosely be described as proto-scientific and other beliefs that are definitely beyond scientific verification.

Chapter 7 examines the intellectual choice we have to make: should we adjust our thought frames to the biblical perspective, or must we reinterpret biblical ideas to be more conformable to those of our own times and culture? I will propose something of each. Science has revised our understanding of the physical world and how it functions – this should unquestionably and unashamedly be regarded as an advance or improvement on earlier knowledge, because we are closer to a true description of how the physical world works. These relatively recent science-based insights will inevitably have some impact on the way we read and understand the Bible. But in other respects we have to adjust our thought frame to the biblical, for three main reasons:

(a) *In some sense* the Bible is foundational for Christian belief (though the exact *sense* has to be explored, and our starting assumptions may be changed as a result of the findings described in this book);

(b) Some aspects of Christian belief which may be assumed or stated in the Bible lie beyond the scope of science to confirm or deny (this issue is explored in Chapter 6);

And (c) the New Testament contains the closest witness we have to Jesus Christ, who is and always must remain the focus and the Lord of the Christian church. So it is inherently imperative that some views and teachings of the New Testament should be brought with urgency to the twenty-first century.

We know many things that the biblical writers did not know. Scientific investigations have made this possible, and we should not be ashamed of nor deny our increased knowledge. This may give the impression that at times I am raising serious and critical questions about the biblical text. I hope that I have already said enough to convince readers that this book is neither anti-Bible nor anti-Christian. The opposite is the case. When making comparisons between biblical views and the more recent understandings informed by science, I am not making anachronistic judgements against the biblical views. A text can only be judged in its contemporary setting. The Bible is central to Christian faith and therefore needs to be well understood. However, present-day Christians can only be confident in their faith while also taking into account present-day knowledge. The Bible was written in times and conditions very different from our own – different in the ideas about the world itself and how it operates, different in the understanding of human life, and different in the perception of the divine. At the very least, our understanding of the Bible will improve if we understand more of the worlds from which it came.

A note on Stories and Myths

Most readers of the Bible recognise that some of its stories have the character of folk-tales or "myths". We must be careful to understand the word "myth" in its technical, literary sense rather than the meaning the word has acquired in much popular usage. In common speech a myth is simply a story which is untrue: for example, "He told me he was at the football match, but that was a myth." He was not at the game, he was elsewhere – he had told a lie. In its technical sense, however, a myth is "an ancient traditional story of gods or heroes, especially one offering an explanation of some fact or phenomenon" (Chambers Dictionary, 1993). So a myth is probably not "true" in the sense that if you had been there with a camera or recorder you would have observed exactly what the story tells – and even the idea that there was a "there" where you might have been with your camera or recorder is dubious. But a myth is much more than an untrue story. It is a story about gods or heroes, as the Chambers' definition states, with a meaning relevant to the times of the author making use of the story. It gives a meaning or an interpretation of events or places within the author's own experience, or it offers insights into the general state of mankind, or the nature of divinity, or the relationships between mankind and God or the gods. If the author's interpretation is accepted by the reader, the myth may still be regarded as "true", though in a sense different from what might be called factual truth.

The uneasiness that some people feel when the language of myth is applied to the Bible relates to the question of where to draw the line. If one story in the Bible is regarded as myth in the sense described in the preceding paragraph, what is to prevent the same ascription to all the stories of the Bible, even those most precious to personal faith and Christian life? If, for example, the stories of creation and the Garden of Eden are understood as mythical, could the Gospel stories of the life of Jesus be treated in the same way? Fears such as these lead some people to apply blanket criteria to all the texts of the Bible, treating them all as factually true as though recorded by a faultless and non-selective observer. In my view, such practices diminish the value and richness that can be achieved from a more subtle approach to the differing character of the biblical material.

Unfortunately, there is no simple answer to this difficulty. The Bible contains no marginal annotation saying "Here be myth", or "This story comes from an ancient tradition", or "The likeness of these characters to any real person living or dead is only approximate". No – it is necessary to read intelligently and to discern the internal signs which distinguish one kind of text from another.

Many such signs are available. First, a simple distinction exists between prose and poetry, and modern translations of the Bible help readers who are unfamiliar with the original languages to identify these differing types of text by printing poetry in verse form. Even without the assistance of poetic structure, it will often be obvious from the language that imaginative description is being used. When the lovers in the Song of Solomon say "Your eyes are doves" or "His legs are alabaster columns" (Song of Solomon 1.15; 4.1; 5.15) we know that we have to apply poetic imagination to the passage – even so, we might struggle to determine how "doves" is a complimentary description of the woman's eyes. Other clues about the nature of a text are found in the style of writing, or the repeated use of particular words, or the theological or political outlook which the text displays. Internal contradictions or interruptions to the train of thought may indicate where texts from different sources have been combined. Even those biblical books which appear to be historical, such as 1 and 2 Samuel and 1 and 2 Kings, must be read in the awareness that all history is written from a particular perspective and with a purpose, which will colour the authors' selection of material and methods of presentation.

In short, there are no short cuts in the intelligent reading of Scripture. The Bible is a library which has come to us through a complex but largely unknown series of events. We should be ready to see a range of literary forms and a variety of points of view. We should read intelligently and openly.

One unavoidable problem raised by an intelligent, scientific approach to the Bible is the nature of "miracles". This theme is considered in more detail in Chapter 5, and without at this point pausing to define an exact meaning of miracle, it can scarcely be denied that the phenomenon of miracle is strongly present in the Bible, in both the Hebrew Scriptures and the New Testament. Does the acceptance of myth as a literary form lead inevitably to questioning whether the miracles described in the Bible should be regarded as actual events?

This is a huge question, with its own extensive literature.[5] The very meaning of "miracle" is not as clear as is sometimes assumed. The scientific world is itself now more ready to accept the occurrence of extreme and abnormal events, inexplicable on the basis of current knowledge, than was the case in, say, the nineteenth century. This is one of the ways in which scientific thought and religious belief have actually drawn closer in their methodologies rather than further apart.

Some religious people would say that the acceptance of miracles is an essential part of belief in God. In his very nature as supreme over the physical world, God is able to perform actions that are impossible for any human and may appear contrary to the physical order. Texts such as the following suggest that the ability to perform miracles is an integral aspect of God's being:

You are great and do wondrous things; you alone are God.
Psalm 86.10
Remember the wonderful works he has done, his miracles, and the judgements he has uttered.
Psalm 105.5
Who alone does great wonders.
Psalm 136.4

Psalm 86.10 may appear to be a statement of monotheism, but it has been suggested that "you alone are God" has here the sense of "matchless" or "unique".[6] However, the sense that God does wonderful things beyond the power of humans (and, possibly, of any other gods) is certainly present. Walter Brueggemann has proposed that "wonders" should be translated more emphatically as "impossibilities", though he recognised that in liturgical usage some of the intensity of this meaning is lost.[7] The couplet "great wonders" in Psalm 136.4 is supported by the Hebrew text, though regarded by some commentators as tautologous.[8] Equally, it may be accepted as emphatic.

The central miracle in this biblical story of miracles, it is frequently claimed, is the resurrection of Jesus Christ from the dead. It is commonly argued that if God can accomplish that, other miracles should not present any intellectual problems. Or, conversely, if other miracles are questioned, the resurrection of Jesus and the entire foundation of the Christian faith are open to doubt. I suggest that we need a more careful approach than these arguments would imply. It is possible to question some reported miracles while affirming others. In this respect the texts of the Hebrew Scriptures and the New Testament are of very different character: the elapsed time between the events reported and their commitment to written text is much shorter in the case of the New Testament, and the availability of supporting witnesses much greater. I believe it is possible to maintain a faith in the resurrection of Jesus while exploring critically other biblical accounts of miracle. In this book I will be particularly concerned to examine those reported miracles which touch upon a scientific understanding of the processes of the natural world.

It is important to note that those parts of the Bible regarded as myths or educational stories do not necessarily inform us of the real views of the biblical writers about the operation of the natural world, though they may contain some elements of the writer's worldview. To give an example, which will anticipate the subsequent identification of the passages which I believe should be regarded as myths or educational stories, the fact that the book of Jonah tells of a large fish which swallowed Jonah whole and three days later spewed him alive and well on a beach should not be taken to mean that the writer really believed that such things actually happen. It is a story, possibly even a children's story, and in such stories impossible things happen. The naive literalness in the way this story has been understood by generations of Bible readers is a sad comment on the absence of literary imagination.

In some cases, the text will contain a mixture of mythical elements alongside the actual views of the author(s) about the real nature of things. The creation narratives of the early chapters of Genesis, for example, include an authentic picture of the ancient Hebrew understanding of the cosmology of the universe (or, perhaps more accurately, *one* Hebrew understanding of cosmology – we should not assume this was the only one in circulation), combined with mythical stories of the primal human ancestors. It should not be supposed that the original author(s) actually believed that snakes could once talk (even though this seems to have been believed by some subsequent commentators!). Most present-day interpreters accept, of course, that these chapters of Genesis include passages which arose from very different literary and theological traditions.

So which parts of the Bible would I exclude as evidence for an ancient view of the natural world because of their mythological or story-telling character? (They should not, of course, be excluded from the study of other theological questions). First, the early chapters of Genesis include many elements which are mythological in this sense: the creation stories, the lives and ages of the patriarchs, the strange story of the union between the sons of God and the daughters of men (Genesis 6.1-4) which has been described as "undisguised mythology" [9], and the story of Noah's ark and the flood. These chapters, purporting to tell the story of the earliest period of human life, clearly contain many elements which fit the dictionary definition of "traditional stories of heroes" – or patriarchs. Some interpreters, I believe, have done great disservice to the relationship between Christian faith, the Bible and scientific knowledge by attempting to create scientific explanations of these events as literally told. These chapters are stories, and the attempt to make them consistent with our present scientific knowledge is misguided.

In addition to these early chapters of Genesis, I would argue that the books of Jonah and Daniel should not be regarded as accounts of historical events, but as educational stories (possibly with a germ of a basis in historical persons or events). They should not be read as descriptions of how the world works.

2

"God is in heaven"

The Lord looks down from heaven.
Psalm 14.2

Where does God live? In this study of the ways in which the cultural assumptions behind the biblical texts differ from the twenty-first century mentality guided by scientific knowledge, I begin with the question of the location of God.

Much of the Bible gives a clear impression that God lives in a quasi-physical location, which exists "upwards" from the earth, in or above the sky. Most Christian believers in our own time will of course reject this primitive idea, but it does our understanding of the Christian faith no good to deny that this view is present in many parts of the Bible. It is tempting to regard these statements about the "upwards" location of God as metaphorical, but when we make the effort to enter the cultural assumptions of the Bible it is difficult to deny that a degree of literalness must exist in these physical descriptions.

These texts must be balanced by others that refer to a more sophisticated understanding of the universal, non-physical presence of God. These other texts will also be examined below.

The idea that God or the gods should be found "upwards" is deeply ingrained in many religions and cultures. The high points on the surface of the earth, hilltops and mountains are frequently given special sacred significance. Why should this be so? It is understandable that gods should be regarded as somewhat remote from the world of human experience, but why should the gods be thought of as "up" rather than "down" or even "across", i.e. horizontally distant but still on the surface of the earth? The sense that gods are different from and superior to humans probably rules out a parallel location somewhere on the earth. The word "superior" itself, meaning "higher than", indicates how deeply ingrained is the idea that higher means better. Upwards is the direction of the sun, which for all peoples has always had such a profound influence on human life. The moon and stars and planets, which in many cultures were also thought of as influencing events on the earth, are also sensed as upwards (though in reality all these heavenly bodies are all around us, and sometimes underneath!). The sky is also the place from which the weather appears to come, so the occurrence of rain, winds and storm, clearly beyond the control of mankind, seems to be determined from above. Downwards, on the other hand, is bound to be dark, because it is permanently shielded from the light of the sun, and is therefore more likely to be regarded as the home of evil rather than good.

The location of God according to the Hebrew Scriptures

Because of this sense that the gods are to be found upwards, ancient peoples tended to locate their temples and shrines at high points in the landscape, at least in those areas with sufficient variations of relief to offer sites on mountains or hills. The Hebrew Scriptures provide plenty of evidence that Canaanite and Baalite shrines, against which the Jewish prophets railed so urgently, were located on hilltops. The meeting between Moses and God, so foundational for much of the religion of Israel, took place significantly on the mountain top of Sinai. And even the location of the Jerusalem temple, on a rather modest hill, is described in exaggerated terms as a mountain top. [10] Hans-Joachim Kraus and other authors associate the idea of God looking down on the whole world with the elevated position of Jerusalem where God is enthroned[11], but Jerusalem could hardly be a vantage point for seeing as far as God does. More probably, the hill-top location of the temple was regarded as a poor copy of the heavenly viewpoint of the "God Most High."

Let us look briefly at some of the many examples of texts in the Hebrew Scriptures which regard God as physically located "up above". The Psalms are particularly rich in this type of language. God is not merely *located* in a high position, but from there he also *sees* all that is happening on the earth, and he *rules* from his throne set in the heavens:

Numerous verses in the Psalms affirm the presence of God in the heavens:

> *He will send from heaven and save me.*
> *Psalm 57.3*

The expectation that God will send help "from heaven" in Psalm 57.3 contrasts with the refugee's shelter "in the shadow of your wings" just two verses earlier. The wings refer metaphorically to the protection a bird gives to its young, and probably also identify the psalm as an expression of refuge in the Temple courts, where God's presence was represented by the outstretched wings of the cherubim in the holiest inner place. This Psalm therefore seems to contain a combination of expectations that God will deliver the suppliant both "from heaven" and through the provision of the Temple refuge for persons unjustly accused.

> *For who in the skies can be compared to the Lord? Who among the heavenly beings is like the Lord, a God feared in the council of the holy ones, great and awesome above all that are around him?*
> *Psalm 89.6-7*

The heavenly council depicted so vividly in Psalm 89.6-7 is also found in the polytheistic texts of other Near Eastern cultures, such as those of the fifteenth century BCE tablets from Ugarit. In the Israelite version, God is seen as the unquestioned ruler who nevertheless seeks the advice of the accompanying beings, and sometimes puts into effect their recommendations. The name "council" is therefore not an empty concept – it represents an advisory body, options are discussed and decisions are reached. For other biblical references see 1 Kings 22.19-23, Psalm 82, Job 1-2 and Jeremiah 23.18. The passage in 1 Kings 22 contains the theologically difficult concept of Yahweh approving the offer of a lying spirit to deceive King Ahab, but the point for our present purpose is that this action was decided in a heavenly consultation. The imagery depicts a place where these discussions were held, "in the skies".

The Lord is exalted, he dwells on high; he filled Zion with justice and righteousness.
Isaiah 33.5

Blenkinsopp comments on this verse that although the idea of God dwelling "above" may have originated at an early stage, it persisted in later liturgical and devotional language: "That God's dwelling is above, in the sky, is an ancient representation, but it occurs in Second Temple texts, no doubt in response to the need to emphasize the transcendent and universal character of the deity."[12] This recognises that at various stages in Israel's history the dwelling of God on high was conceived both literally and metaphorically. The second half of Isaiah 33.5 shows how God's exaltation in the heavens was associated with his presence in the temple in Jerusalem.

From this heavenly vantage point, God *sees* all events on the earth:

The Lord looks down from heaven on humankind
Psalm 14.2; also Psalm 53.2[13]*; 80.14; 102.19*
The Lord is high above all nations, and his glory above the heavens.
Who is like the Lord our God, who is seated on high, who looks far

down on the heavens and the earth? Psalm 113.4-6
though the Lord is high, he regards the lowly; but the haughty he perceives from far away.
Psalm 138.6

Psalm 138 may be a royal song of thanksgiving. [14] Therefore "lowly" may here be a relative description. The king of Israel, though surrounded by the symbols of monarchy, knows that he is a subject of Yahweh, and is especially in danger when he goes into battle (v7). As I have noted in Chapter 1, the nation of Israel-Judah may indeed have felt itself lowly in comparison with some of the surrounding nations. The king of Israel may have taken strength from the belief that though he was the head of a small and impoverished state he was protected by an invincible God.

Not only does God see events on the earth, but God also *rules* from his throne in the heavens. The fact that God is depicted as sitting on his throne indicates the ease and majesty and unquestionable nature of his rule – he does not have to stand or threaten or exert force to exercise his will:

He who sits in the heavens laughs; the Lord has them [the rulers of the earth] in derision.
Psalm 2.4[15]

Some of the psalms expressing the kingly rule of God are interpreted as royal psalms or even coronation psalms – the rule of God was celebrated in the context of the enthronement of the Israelite king. The celebration often had a fervent nationalistic tone – the God of Israel and the king of Israel together were an impregnable match for any neighbouring powers. Any monarchs or dignitaries from nearby states who were present at the coronation would be left in no doubt about the message conveyed to them. For example, Psalm 2 contains four stanzas:

Verses 1-3:	The psalm mocks those other nations, peoples and kings who oppose Israel's God and Israel's king.
Verses 4-6:	The enthroned Lord himself laughs in derision at them, and declares that the king enthroned on Zion is indeed a divine appointment.
Verses 7-9:	The king states the terms of his appointment, possibly reading from the "terms of appointment" given to him – see 2 Kings 11.12 for the existence of such a document or inscription, a "form of 'protocol' or list of regulations governing the conduct of kingship in Israel and Judah." [16]
Verses 10-12:	The psalm exhorts neighbouring kings to show suitable deference to Israel's king and God.

This association between the coronation of the king and the celebration of divine kingship is a mutual interaction of reality and metaphor. The ceremonies of kingship and the claims of authority on the part of the king of Israel provide imagery for the rule of God, while at the same time the king derives his own role from the majesty of God who is already enthroned over the world. Israel was not alone among the nations in making this connection between their king and their god.

The Lord sits enthroned for ever, he has established his throne for judgement.
Psalm 9.7

Psalm 9 has an acrostic pattern which runs on into Psalm 10. The royal rule of God is, in this case, not so clearly located in a coronation setting, though the extent of God's rule over the "enemies" and the "nations" is apparent (9.3-8, 15, 19-20; 10.16).

The Lord is in his holy temple; the Lord's throne is in heaven. His eyes behold, his gaze examines humankind.
Psalm 11.4

Psalm 11.4 brings into immediate juxtaposition the two poles of Israelite worship: the Lord is in heaven from where he examines the whole of humankind, and also in his temple where Israel meets him in worship. As Hans-Joachim Kraus expresses it, statements about God's residence on Zion and his exalted reign and judgment of the entire world "lie immediately next to each other and among each other".[17]

The Lord sits enthroned above the flood [i.e. above the storehouses of water in the sky]; the Lord sits enthroned as king for ever.
Psalm 29.10

Psalm 29, with its recognition of the place of God amongst the heavenly council or "sons of God" (v1), and its affirmation of the rule of God over the waters (v3), the flood (v10), and the cedars of Lebanon (v5), and displayed in thunder (v3), earthquake (vv 6, 8) and storm (vv 7-9), is sometimes regarded as an adoption from a Canaanite or Baalite liturgy. The enthronement of God over the flood expresses his rule over the forces of chaos which might still resist his rule, and specifically the replacement of Baal by Yahweh as the ruler of the forces at work in events in the natural world.

The Lord looks down from heaven; he sees all mankind. From where he sits enthroned he watches all the inhabitants of the earth.
Psalm 33.13-14

Ascribe power to God, whose majesty is over Israel; and whose power is in the skies.
Psalm 68.34

The Lord has established his throne in the heavens, and his kingdom rules over all.
Psalm 103.19

Our God is in the heavens; he does whatever he pleases.
Psalm 115.3

This kingly aspect of divine activity may be compared with the behaviour of the deities described in the tablets from Ugarit, who sometimes competed for a suitable throne in the heavenly mountain. The senior god El was recognised as supreme in the Ugaritic pantheon, but conflicts occurred between the junior gods Baal, Yam (the sea) and Mot (death). At one point a young god Athtar attempted to occupy Baal's throne but was found to be too small and had to renounce this position.[18] Some of the biblical Psalms of kingship may borrow from such sources, but the rule of God which they present has been cleansed of most of the polytheistic conflicts and relativities present in the Ugaritic system – though Psalm 82.1 still describes God's place in the heavenly council as "in the midst of the gods" and Psalm 138.1 offers praise to God "before the gods".

Although such references to God's dwelling in or above the skies are particularly frequent in the Psalms, they are by no means limited to that source. The later part of the book of Isaiah describes God as "he who sits above the circle of the earth, and its inhabitants are like grasshoppers; who stretches out the heavens like a curtain, and spreads them like a tent to live in" (Isaiah 40.22). Later chapters of Isaiah take up the same theme:

Thus says the high and lofty one who inhabits eternity, whose name is Holy: I dwell in the high and holy place.
Isaiah 57.15

Such fasting as you do today will not make your voice heard on high.
Isaiah 58.4

Look down from heaven and see, from your holy and glorious habitation.
Isaiah 63.15 (also Psalm 80.14)

O that you would tear the heavens and come down.
Isaiah 64.1

Heaven is my throne and the earth is my footstool.
Isaiah 66.1

These chapters of the book of Isaiah probably date from the period immediately following the return of Israel from their captivity in Babylon. The prophet has a positive view of the rebuilt temple in Jerusalem (see Isaiah 56.4-7; 57.13; 60.7, 13; 65.11-12) but does not subscribe to a naive view that the temple is the place where God "lives".[19] The prophet wished to point towards God "on high" rather than being encountered in any building or ritual. Heaven as God's throne and the earth as his footstool is contrasted with any "house that you would build for me" because all things have been created by him and therefore belong to him (66.1-2). Blenkinsopp comments that the change in emphasis from God resident in the temple to God on high "reflects a shift in thinking to which the destruction of the [previous] temple must have contributed"[20] although, as we have seen above, the location of God in the heavens was also an early theme. The loss of the temple forced Israel to reconsider its temple-oriented faith, and revert to earlier forms of belief.

Isaiah 66 may appear to be a total rejection of temple worship with its sacrifices and other rituals:

Whoever slaughters an ox is like one who kills a human being;

whoever sacrifices a lamb, like one who breaks a dog's neck;
whoever presents a grain-offering, like one who offers swine's blood;
whoever makes a memorial offering of frankincense, like one who blesses an idol.
Isaiah 66.3

However, such a conclusion seems inconsistent with other favourable references to God's holy mountain in these chapters of Isaiah (56.7; 57.13; 65.11 – though the possibility of editorial revisions has to be allowed). The view that the temple was literally the home of God was rejected, but the preservation of national social and religious identity demanded the preservation of a focus on Jerusalem. This was found by locating the "Name" or "Glory" of Yahweh in the temple, "a kind of hypostasis of the deity". These were "attempts to express divine transcendence without surrendering immanence." [21]

Other prophets also have the same understanding of the place of God up above in the heavens. The book of Jonah describes God as "the Lord, the God of heaven, who made the sea and the dry land" (Jonah 1.9). The prophet Micah says that "the Lord is coming out of his place, and will come down and tread upon the high places of the earth" (Micah 1.3). And Zechariah says that God "has roused himself from his holy dwelling" and has come to dwell in the midst of Judah and Jerusalem (Zechariah 2.10-13). The tone of these statements by Micah and Zechariah suggests that God will emerge from his normally invisible location and will do things that are visible and active.

The historical books display the same view of a God who is enthroned above the earth. In the book of Genesis we read that the Lord "came down" to see what was happening on the earth (Genesis 11.5, 7; 18.21)[22], or "went up" after speaking to Jacob (35.13). The Lord also "came down" to Mount Sinai to meet with Moses (Exodus 19.11, 18, 20; 34.5). The Deuteronomist instructs the people to ask God to "look down from your holy habitation, from heaven, and bless your people" (Deuteronomy 26.15). The prophet Elijah is depicted at the end of his life as having "ascended" into heaven in a chariot of fire and a whirlwind, until he could no longer be seen by the watching Elisha (2 Kings 2.11-12). The Chronicler describes the Jerusalem temple as the footstool of God (1 Chronicles 28.2), a variant on the whole earth as the footstool of God as we have seen in the case of Isaiah 66.1. King Jehoshaphat asks, "Are you not God in heaven? Do you not rule over all the kingdoms of the nations? In your hand are power and might, so that no one is able to withstand you" (2 Chronicles 20.6). In the time of King Hezekiah, the prayer of the priests and Levites "came to [God's] holy dwelling in heaven" (2 Chronicles 30.27). Nehemiah prays to "the God of heaven" (Nehemiah 1.4-5; 2.4, 20).

Finally, the wisdom writers also use the same language about God's dwelling place "high" in heaven. "Is not God high in the heavens? ... he walks on the dome of heaven" (Job 22.12-14), that is, God lives above the heavenly firmament. "Dominion and fear are with God; he makes peace in his high heaven" (Job 25.2), that is, he affirms his authority over any heavenly opposition. Last but by no means least, "God is in heaven, and you upon the earth; therefore let your words be few" (Ecclesiastes 5.2). The distance between God and the earth is here used to discourage the arrogant use of the tongue.

This accumulation of material from different parts of the Hebrew Scriptures gives a clear picture of God residing high above the earth, in a location described as heaven or the heavens, from where God both sees and rules over what is happening on the earth. Nevertheless, this universal picture has to be modified in three ways.

First, although God's dwelling place in heaven is described as a physical location, it is not regarded as a *single* location. Even in terms of the Hebrew cosmology of a flat earth surmounted by a dome-like "firmament", heaven could not be allocated a specific position. It could not be described as "there" but not at some other position in the heavens. To use modern topographical terminology, it could not be located by x, y and z three-dimensional coordinates, nor by the declination and right ascension as when describing the position of a star observed from the earth. From his position in heaven, God can see everything and rule everything. God's location appears to extend everywhere throughout the heavens. While physical language is used of God's high home in heaven, it is not limiting as a truly physical location would be. The high location of heaven as described in these biblical texts is better described as quasi-physical rather than strictly physical.

Second, we must recognise that according to the Hebrew Scriptures the location of God in heaven does not prevent God from also being present on the earth. Although the texts describing God's presence on the earth are fewer than those referring to his home in heaven, they are nevertheless present. The whole story of the Hebrew Bible is an account of God's actions with his people on the earth, even though in many places the precise location of God when he acts is not discussed.

Thus, when the Deuteronomist states that "the Lord is God in heaven", consistently with the view we have been examining above, he adds "and on the earth beneath" (Deuteronomy 4.39). Similarly, he writes that "heaven and the heaven of heavens belong to the Lord your God", and also "the earth with all that is in it"; and out of this abundant wealth, God especially chose Israel for his purposes (Deuteronomy 10.14-15). The commandments which God gave are not remote in heaven, nor do they have to be sought across the sea as the Mesopotamian hero Gilgamesh had to search, but they are near, "in your mouth and in your heart", that is to say spoken and remembered (Deuteronomy 30.11-14). In his prayer to God at the dedication of the Jerusalem temple, King Solomon recognises that "heaven and the highest heaven cannot contain you", though, interestingly, God still hears from "heaven your dwelling place" (1 Kings 8.27, 30; cf. 2 Chronicles 2.6; 6.18, 21). Much of the Temple liturgy relied upon subtle parallels between the presence of God in heaven, in the Temple itself, and in the nation. In the language of later theology, this has been described as the "classic theological tension between the immanence and transcendence of God".[23] However, a qualification of this view should be entered on two grounds: first, both immanence and transcendence are normally used as abstract, non-spatial descriptors of God's being, thus missing the quasi-physical sense in the Hebrew Scriptures; and second, the perceived balance between these ideas was not the same in the times of the Jerusalem Temple as in our own.

The universal presence of God is also recognised in the Psalms: "you are near, O Lord" (Psalm 119.151); and "Where can I go from your spirit? ... If I ascend to heaven ... if I make my bed in Sheol ... if I take the wings of the morning [i.e. eastwards] and settle at the farthest limits of the sea [i.e. westwards] even there your hand shall lead me" (Psalm 139.7-10).

Among the prophets Jeremiah, speaking for God, asks rhetorically "Do I not fill heaven and earth?" (Jeremiah 23.24). And the prophet Amos warns that even in Sheol there will be no escape from God:

Though they dig into Sheol, from there shall my hand take them;
though they climb up to heaven, from there I will bring them down.
Though they hide themselves on the top of Carmel ...
and though they hide from my sight at the bottom of the sea ...
And though they go into captivity ...
Amos 9.2-4

The universal presence of God in heaven and in Sheol is also found in some texts from El Amarna.[24]

A *third* modification to the apparent physicality of God's home in heaven must also be recognised. Within physical reality, a location within physical space also implies a position within a historic timeframe. Even without Einstein's insight into the continuity of space-time, it could be recognised that a person, thing or event located in a specific place is also conditioned by time or the passage of time. In the case of the Hebrew understanding of God, however, the quasi-physical location of God in heaven carried no such limitation of God to a particular timeframe. God was understood to be in some way timeless, even though he shared in the time-related experiences of human beings. Thus, "the Lord sits enthroned for ever" (Psalm 9.7; 29.10), God is "enthroned from of old" (Psalm 55.19; 74.12; 93.2; Habakkuk 1.12), and "a thousand years in your sight are like yesterday when it is past, or like a watch in the night" (Psalm 90.4). As already noted, God "inhabits eternity" (Isaiah 57.15). An expression such as "for ever" does not necessarily carry the sense of infinity or timelessness as understood by modern mathematicians or cosmologists – rather, it means as far back in time as it is possible to imagine, and then more: from everlasting in the past to everlasting in the future (Psalm 90.2).

The downwards alternative - Sheol

A corollary to this upward direction towards God is the Hebrew understanding of the underworld, usually described as Sheol or the Pit. If God is thought of as upwards, the spirits of humans, or at least the spirits of wicked humans, appear destined for a downwards journey to an undesirable, shadowy existence beyond the grave. References occur particularly in the books of Job, Proverbs and in many Psalms, and this belief may be a late development in Hebrew theology.

A gruesome satire on the unpleasantness of Sheol is found in Isaiah 14, where the prophet celebrates the arrival there of the defeated King of Babylon. Other defeated kings of the nations can hardly stir themselves from their miserable existence to greet the newcomer:

Sheol beneath is stirred up to meet you when you come;
it rouses the shades to greet you, all who were leaders of the earth;
it raises from their thrones all who were kings of the nations.
All of them will speak and say to you:
'You too have become as weak as we are!
You have become like us!'
Your pomp is brought down to Sheol, and the sound of your harps;
maggots are the bed beneath you, and worms are your covering.
Isaiah 14.9-11

In the above case, Sheol is seen as the destiny of foreign kings, perhaps specifically those who have opposed Israel. More generally, the question "Who goes to Sheol?" was answered in several ways.

In some cases, Sheol appears to be the inevitable destination of all: for example, "Who can live and never see death? Who can escape the power of Sheol?" (Psalm 89.48), and "Whatever your hand finds to do, do with your might; for there is no work or thought or knowledge in Sheol, to which you are going" (Ecclesiastes 9.10). In other verses, it appears to be the wicked or those of other nations who are threatened with this fate, for example: "Drought and heat snatch away the snow-waters; so does Sheol those who have sinned" (Job 24.19), and "Do not let me be put to shame, O Lord, for I call on you; let the wicked be put to shame; let them go dumbfounded to Sheol" (Psalm 31.17). Similarly, in an extended passage in Ezekiel, it is the kings and soldiers of the nations opposing Israel who are found in Sheol (Ezekiel 32.17-32), as in Isaiah 14 quoted above. A yet further view is that the wise may be able to avoid Sheol: "For the wise, the path of life leads upwards, in order to avoid Sheol below" (Proverbs 15.24). The words "upwards" and "below" were not included in the Septuagint (LXX) translation from Hebrew into Greek, placing the emphasis on a satisfying this-worldly experience.[25] However, the reference to Sheol or Hades is retained in the LXX, thus suggesting that an experience after death is still in view. On the other hand Roland Murphy writes that upward and downward "have nothing to do with a life beyond death" but refer to "upwards into life" or "downwards to Sheol, which can be understood literally as death or metaphorically as adversity."[26] Against this view, it is difficult to avoid the impression that Sheol usually refers to something beyond death, however low-grade that experience might be. For our present interests, the possibility that according to this text the wise may be able to avoid Sheol remains present.

Psalm 6.4-5 offers yet another perspective. "Save my life" pleads the psalmist, "for in death there is no remembrance of you; in Sheol who can give you praise?" If he enters Sheol he will be unable to praise God – is there even a hint that God will be the loser if he cannot receive the psalmist's praise?

A yet further understanding of the place of Sheol is present in the book of Job: Sheol may be inevitable but nevertheless offers some prospect of hope:

> *Mortals lie down and do not rise again;*
> *until the heavens are no more, they will not awake or be roused out of their sleep.*
> *O that you would hide me in Sheol, that you would conceal me until your wrath is past, that you would appoint a set time, and remember me!*
> *Job 14.12-13*

These verses open up many interesting thoughts: Job is pleading that God will shelter him from God's own wrath, even though the only prospect is a shelter in the otherwise undesirable Sheol. This text envisages the possibility that God may have competing purposes, even conflicting purposes. Sheol is not a pleasant place, but even so it may be a refuge from something worse, until better times appear.

The location of God according to the New Testament

When we turn to the Christian writings of the New Testament we encounter a similar duality between the presence of God in heaven, which is described in quasi-physical terms in an upwards direction, and the presence of God in events on the earth and in human lives.

The Jewish character of Matthew's gospel is well known, so it is not surprising that in this gospel the sayings of Jesus about God are expressed in terms very similar to those of the Hebrew Scriptures. "Whoever swears by heaven, swears by the throne of God and by the one who is seated upon it" (Matthew 23.22). Heaven is the throne of God, and the earth his footstool (5.34-35), an image which contains a clear upwards and downwards dimension and is very reminiscent of verses already noted in Isaiah 66.1 and 1 Chronicles 28.2. Matthew's descriptions of God as "Father in heaven" (5.45; 6.1, 9; 7.11, 21; 10.32-33; 18.10, 14, 19) do not explicitly contain an "upwards" orientation, but when seen against the background of the Hebrew Scriptures they must surely be understood in this way. The reference to children's "angels" continually seeing the face of the Father in heaven (18.10) suggests a strong locational sense in a place other than the earth. The earth and heaven are contrasted in a quasi-locational manner in the Lord's Prayer at Matthew 6.10 ("on earth as it is in heaven"), in the saying about treasures at 6.19-20 ("treasures on earth, where moth … rust … thieves ... , … treasures in heaven, where neither moth nor rust … nor thieves ..."), and in the obscure saying at 18.18 concerning binding and loosing ("whatever you bind on earth will be bound in heaven, and whatever you loose on earth will be loosed in heaven"). These verses in Matthew's gospel provide clues about Jesus' perception of events in heaven and therefore ("on earth as it is in heaven") how things should be on the earth.

The gospels of Mark and Luke also describe the vertical orientation of heaven, in stating that Jesus was "taken up" or "carried up" into heaven (Mark 16.19, the probably non-original longer ending of this gospel; cf. Luke 24.51). The New International Commentary notes that "The abode of God, heaven, was 'up', so, in the 'universe' of his day, one went 'up' to meet God. In this account, movement 'upward' signifies in a visible and concrete way the elevated status of Jesus."[27]. Howard Marshall understands the imperfect tense in Luke 24.51 as indicative of "a gradual departure, as in Acts 1.9f."[28] Luke's account of the ascension in the Book of Acts includes even more detail: "he was lifted up, and a cloud took him out of their sight", and while they were "gazing up towards heaven" two men dressed in white robes told them that Jesus "who has been taken up from you into heaven, will come in the same way as you saw him go into heaven" (Acts 1.9-11). The literalness of Luke's understanding of an upward ascension is accepted by James Dunn: "It was no problem for Luke, of course, since he presumably shared the common perception of the cosmos, where heaven was literally above the earth, so that 'going to heaven' meant quite literally 'ascension'."[29] A little later in the book of Acts Peter preaches that Jesus "must remain in heaven until the time of universal restoration that God announced" (Acts 3.21); Christ's departure into heaven represents a waiting period until the time is right for God's consummation of all things.

Even in John's gospel, which generally has a less locational sense of the distinctions between this world and the presence of God, Jesus emphasises an upwards location for God when he says "You are from below, I am from above" (8.21-23; cf. 3.31; 7.33-36; 16.5). In these typically Johannine verses Jesus talks simultaneously of his imminent death and his heavenly commission and destination. It is also in this gospel that the risen Jesus says to Mary, "I am ascending to my Father and your Father, to my God and your God" (20.17). John's complex treatment of above/below, this world/the world of the Father, has many layers and cannot be reduced to a physical up/down –

nevertheless the language of physical locality is used to make these distinctions.

The Pauline letters also include numerous references indicating a "vertical" view of the place of God in heaven. In the eschatological material in 1 Thessalonians, generally regarded as Paul's earliest extant letter, he writes that Jesus "will descend from heaven" and that those alive at the time "will be caught up in the clouds together with them [the resurrected dead] to meet the Lord in the air" (1 Thessalonians 4.16-17). Against this background, other references in the letter to Jesus coming "from heaven" may also be understood in terms of descent, even though that is not explicitly stated (1.10; cf. 3.13). 2 Thessalonians also refers to the future time "when the Lord Jesus is revealed from heaven with his mighty angels" (2 Thessalonians 1.7).

In 1 Corinthians, Jesus is described as "the man from heaven" or "of heaven" (15.47-49), although again a vertical sense is not explicit here. The idea of a man present in heaven or coming from heaven is found in the vision of Daniel 7.13 and also in Jewish inter-testamental literature such as 1 Enoch and 4 Ezra; Paul found these images readily available to be identified with the Messiah. [30]

According to the letter to the Ephesians, God raised Jesus from the dead "and seated him at his right hand in the heavenly places, far above all rule and authority and power and dominion" (Ephesians 1.20-21). This quasi-physical location of the "heavenly places" (or simply "the heavenlies") may also be inferred in 1.3, 10; 2.6; and 3.10. In discussing "the heavenlies", Andrew Lincoln writes:

> The most appropriate meaning for this phrase in each of its five usages in Ephesians is a local one ... This cosmological picture was one in which heaven was depicted as the upper part of the cosmos but at the same time stood analogically for the spiritual world above, which the firmament concealed,

> and beyond that referred also to the realm of divine transcendence ... for the early Christians Christ has not simply disappeared nor had he evaporated into a universal spirit, but he had departed to a new sphere, that of heaven, which would be appropriate to his transformed body's mode of existence.[31]

Ephesians 2.2 also seems to be another significant verse in affirming the cosmic nature of the location of God and the salvific achievements of Christ: the enemy whom the Christian way is confronting is described as "the ruler of the power of the air". The opposition encountered by the followers of Christ is strongly personalised, and dwells in the air, that is the same realm as inhabited by humans and immediately above.

The descent of Christ from heaven is also indicated in the letter to the Philippi: "Our citizenship is in heaven, and it is from there that we are expecting a Saviour, the Lord Jesus Christ" (3.20). The well-known hymn to Christ in Philippians 2 also has a descent-ascent theme: he emptied himself, taking the form of a slave, was born in human likeness and became obedient to death; "therefore God has highly exalted him ... so that ... every knee should bend, in heaven and on earth and under the earth" (2.7-10). In an ethical exhortation to the church at Colossae, the writer encourages his readers to "seek the things that are above, where Christ is, seated at the right hand of God" (Colossians 3.1f).

The same kind of "vertical" description of the location of heaven is found in the other books of the New Testament. The Letter to the Hebrews writes in Pauline-type language of Jesus "seated at the right hand of the throne of the Majesty in the heavens" (1.3; 8.1; 12.2; cf. Psalm 110.1 and Zechariah 6.13 for the background in the Hebrew Scriptures). Again according to the Letter to the Hebrews, Christ has "passed through the heavens" (4.14) and is "exalted above the heavens" (7.26), probably a reflection of belief in a layered heaven as the abode of spiritual beings, with the presence of God at the most exalted level. In 8.4

his present heavenly priestly role is contrasted with what he would be "if he were on earth". The epistle of James states that every perfect gift is "from above, coming down" from the Father of lights (1.17); "wisdom from above" may be contrasted with earthly, unspiritual wisdom which "does not come down from above" (3.15, 17). The first epistle of Peter states that Christ "has gone into heaven and is at the right hand of God, with angels, authorities and powers made subject to him" (3.22).

Finally in this examination of the biblical material, the book of Revelation includes a series of visions involving events in heaven above and the earth below, and exchanges between these two realms. Christ is depicted as "coming with the clouds" (1.7, cf. Matthew 24.30), drawing on an image in Daniel 7.13, itself derived from an ancient characteristic of the Canaanite god Baal. John sees a door in heaven, and is invited to "Come up here" (4.1; also 11.12, where the martyrs "went up to heaven in a cloud"). In the opposite direction, an angel came down from heaven wrapped in a cloud (10.1); the association with clouds has been taken to symbolise the divine presence, as in parts of the Hebrew Scriptures[32], while it also conveys a locational sense upwards or in the sky. Similarly the dragon is "thrown down" from heaven to earth (12.9-13; cf. the saying of Christ at Luke 10.18, "I saw Satan fall like lightning from heaven"), and a beast makes fire "come down from heaven to earth" (13.13; cf. 20.9), reminiscent of actions by Elijah at 2 Kings 1.10-12. As in the Philippians hymn, heaven and earth are described as a hierarchy of levels – Revelation adds under the earth and the sea as additional dimensions (5.13). As a glorious conclusion to the visions, it is the heavenly city which John sees "coming down out of heaven from God" (21.2, 10). As Mounce comments: "The descent is an announcement in visionary terms of a future event which will usher in the eternal state. That the city comes down from God means that the eternal blessedness is not an achievement of man but a gift from God." [33] And Witherington writes: "The final state is represented as transpiring not when believers go up to heaven but when God and heaven come down permanently." [34]

Just as the Hebrew Scriptures contain a minority of verses depicting the universal presence of God alongside the "God is in heaven" theme, so also the New Testament includes the recognition of God's presence in an intimate, immanent, sense. Jesus taught that the kingdom of God should be thought of not as something remote in distance or time but near at hand[35] (for example, Matthew 4.17; 12.28; cf. John the Baptist's teaching at 3.2), though his comparisons of the kingdom to a heavenly banquet also preserved a more remote, eschatological view (for example, Mark 14.25). Paul's expression "in Christ" for the present experience of the Christian communities is also important evidence for the first Christians' experience of God on the earth – an immediate awareness of the risen Lord. On the other hand, the experience of the Holy Spirit in the first churches may have held particular importance because the Spirit provided the necessary connection between believers on the earth and God and Christ enthroned in heaven. The ascended Christ, as we have seen, was thought of as enthroned alongside God in heaven, and in some preaching at least, Christ "must remain in heaven until the time of universal restoration" (Acts 3.21). Hence the importance of the Spirit as bringing the presence of God and Christ into intimate reality. Also in the Book of Acts we find Paul preaching to Athenian philosophers that "he is not far from each one of us" and quoting favourably from their own authors, that "in him we live and move and have our being" (Acts 17.27-28). [36]

The New Testament is also similar to the Hebrew Scriptures in that however physical or quasi-physical its concepts of the home of God it sees no limitations of God in relation to time. God is eternal, thus guaranteeing continuity between the messages of his prophets in the past and the gospel of Christ (Romans 16.26). The gospels and epistles include many discussions of the route to eternal life. Whether the New Testament notion of eternity should be thought of as timelessness or the unending continuation of time is a matter for debate[37], but it is clear that the Christian hope in God continues "for ever". In fact the apostle Paul links the idea of God's home in heaven with the concept of eternity:

For we know that if the earthly tent we live in [i.e. the human body] is destroyed, we have a building from God, a house not made with hands, eternal in the heavens.
2 Corinthians 5.1

Modern views of the location of God

The above examples from both the Hebrew Scriptures and the New Testament have shown a consistent picture of God as located in an "upwards" direction in the heavens. The much-maligned Victorian hymn, "There's a friend for little children, above the bright blue sky", had plenty of biblical precedent. But what is a twenty-first century person of scientific inclination to make of all this upward thinking about God in heaven?

Astronomy during the past four hundred years and space exploration during the last fifty years have surely eliminated any possibility that heaven is a place beyond the sky. Cosmological theory indicates that the expanding universe is not expanding into empty "space", which could conceivably be an undetectable location of the home of God, but is expanding into nothing. The limit of the universe is itself the limit of space-time. We understand that God does not require a physical place or location in which to live. God is spiritual, as described by the minority of biblical references which regard God as present everywhere, at all times. God beyond and outside ourselves should be complemented by a perception of God's nearness and inwardness in human consciousness. God cannot be located by spatial coordinates, and the language which speaks to God, or prays to God, in an "upwards" direction is, at best, metaphorical or traditional. As long as it is clearly understood in this way, the language describing an upwards direction of God may be acceptable as a description of God. But frequently the quasi-physical language may be taken too literally, and in this respect some normal forms of worship and liturgy that focus on the "upward" abode of God are in need of serious revision.

3

God and the natural world

In this chapter I examine the way biblical writers understood or interpreted the relationship of God to the non-human natural world. As already noted, it is difficult to separate precisely the biblical interpretation of the human and non-human worlds, but I will look initially at those aspects which are less related to human life. The issues to be addressed will include the explanation of natural features ranging from mountains to bird migration, the role of God in everyday events such as the weather, and in rarer events such as earthquakes and droughts. The relationship of God to agricultural harvests begins to touch upon more human interests, but will nevertheless be included in this chapter. The perception of God as the cause of more fully human disasters such as plagues, deafness, blindness and other illnesses will be reserved until Chapter 4.

Universal or specific explanations?

One important question to be addressed is, "If God is deemed to be active in one particular incident such as a drought or earthquake, is it necessarily the case that God must be similarly responsible in every drought or earthquake?" If a Hebrew prophet, for example, claimed that a particular calamity was an action by God in response to the wickedness of the people, is it justifiable to extrapolate that understanding and to say that all such calamities are directly authored by God for similar reasons? We must be careful not to generate a universal philosophy on the basis of specific incidents. However, if general statements about the activity of God are expressed in the more philosophical parts of the Bible (mainly the wisdom literature of the Hebrew Scriptures) we must also take account of that view. Expressed in modern conceptual language, this question asks whether God is immanently present in every example of a particular type, or does God act occasionally to implement his will?

The work of E. C. Rust

An earlier study of the Hebrew view of the natural world was published by the Baptist scholar E. C. Rust in the nineteen-fifties. [38] Rust recognised that the biblical views of cosmology and the animal world were not binding on present-day believers: "The Biblical world-view must be corrected by the discoveries of modern science, and a rational critique of it should have complete freedom."[39] Rust also wished to maintain, as I do, that some aspects of the biblical view of the world are of permanent truth and value: "There are, however, certain truths and categories which modern science cannot and has no right to challenge. These are the categories which are bound up not with the structure of the world so much as with the relation of God to the world."[40] It seems to me, however, that as his work developed, Rust did not distinguish with sufficient clarity between those parts of the biblical worldview which later Christians are entitled to change, and those parts which are of lasting significance. Rust attempted to distinguish between the Christian understanding, through revelation, of the "metaphysical realm" and what he describes here as "the structure of the world", but he did not carry through this distinction in his examination of particular aspects of the natural world. Perhaps science has moved forward so much, even in the half century since Rust's book was published, that we now expect material explanations of issues which he was prepared to leave as metaphysical mysteries.

Thus, Rust describes the biblical view of events in the natural world as intimately connected both to the life of God and to the righteous or unrighteous behaviour of the people. There is, he claims, no distinction in the Bible between the natural and the supernatural, for the whole of existence is always dependent upon God's continuing creative acts. Although the Hebrew farmer knew that harvests and fruits were the result of both the inherent properties in the plants and hard human labour, yet these things were themselves the gift of God and would be fruitless without God's continuing goodwill.[41] The sinfulness of Adam, the typical man, leads to a curse upon the ground and consequent hard labour (Genesis 3.17-19). If the people do not follow God's commands the land will vomit them out (Leviticus 20.22). And conversely, when the people return to God the land will be well-watered and fruitful (Isaiah 41.17-20). Rust therefore writes of the Hebrew view of a "psychic" life in the natural world, which is intimately associated with the creativity of God and the behaviour of the people.

Rust seems to conclude that this psychic life is a feature of the biblical view of nature that remains valid for modern readers. He wrote that a religious person, "because of his preoccupation with God, tends to ignore and discount the functioning of second causes, and to view all events as the direct expression of the divine will ... second causes will drop into the background .... We must admit the presence of second causes, but the Biblical approach constrains us to admit the direct relation of every event to the sustaining power of the divine will."[42]

I find this distinction between primary and secondary causes not particularly helpful or illuminating. At what point does one level of "cause" take precedence over or cancel out or merge into another level? In fact, they are not "causes" in the same sense, and therefore it seems inappropriate to use the same word to describe such different ideas. God's action must be different in kind from the action of physical and biological forces which determine outcomes in the natural world. This distinction between the metaphysical and the physical levels of understanding was not well developed in the biblical world. Rust's book recognises that the distinction exists, but does not pursue it as a necessary part of scientific understanding. In this chapter I will develop a more distinctive separation between the biblical and the scientific understanding of the natural world.

In a recent study, Noah Efron draws attention to the inconsistent treatment of the natural world in the Hebrew Scriptures. [43] There was little interest in the regular movements of the heavens, in contrast to the detailed astronomical tabulations of the Babylonians. Natural events on the earth were regarded, at least in some examples, as "portentous" or religiously significant, and the classification of animals was important because of the stringency of Israel's food laws. The natural world was seen as both the creation of God and therefore sacred and unknowable, but also open to human exploitation. As we shall see in chapter 5, some attempted to understand nature through divination or astrology, while others forbade such practices.

God and the landscape (mountains and hills, earthquakes, the sea)

In the biblical view, the landscape was almost unchanging, and had been created directly by God. He was the creator of both the earth and the heavens. "By the word of the Lord the heavens were made" (Psalm 33.6) expresses the ancient view of the creativity existing in the divine word. "By your strength you established the mountains ... you silence the roaring of the seas ... you make the gateways of the morning and the evening shout for joy ... you provide the people with grain" (Psalm 65.6-9). The earth, "which he has founded for ever" (Psalm 78.69) seemed from the perspective of human lifetimes, to be permanent.[44] The mountains and hills could be described, in places, as eternal and everlasting:

> *The blessings of your father are stronger than the blessings of the eternal mountains, the bounties of the everlasting hills.*
> *Genesis 49.26*

Hear you mountains, the controversy of the Lord, and you enduring foundations of the earth.
Micah 6.2

In an apparent oxymoron the prophet Habakkuk, however, can recognise that this permanence is conditional upon the exercise by God of his power:

He stopped and shook the earth; he looked and made the mountains tremble. The eternal mountains were shattered; along his ancient pathways the everlasting hills sank low.
Habakkuk 3.6

Earthquakes were not an uncommon event in Israel, as the region lies close to major fault lines and rift valleys. Earthquakes were a reminder that even the solid earth could be shaken and moved by the will of God:

The earth reeled and rocked; the foundations also of the mountains trembled and quaked, because he was angry.
Psalm 18.7 (This psalm is also found at 2 Samuel 22)

O God, you have rejected us, broken our defences; you have been angry; now restore us! You have caused the land to quake; you have torn it open; repair the cracks in it, for it is tottering.
Psalm 60.1-2

The earth quaked, the heavens poured down rain at the presence of God, the God of Sinai.
Psalm 68.8 (cf. Judges 5.4-5)

The mountains melt like wax before the Lord, before the Lord of all the earth.
Psalm 97.5

The anger of the Lord was kindled against his people, and he stretched out his hand against them and struck them; the mountains quaked, and their corpses were like refuse in the streets.
Isaiah 5.25 (see also Isaiah 13.13; 24.18-20; 29.6; 64.1-3)

At his wrath the earth quakes, and the nations cannot endure his indignation.
Jeremiah 10.10

The Lord, the God of hosts, he who touches the earth and it melts.
Amos 9.5 (cf. Micah 1.4; Nahum 1.5)

In a specific event, when Saul's son Jonathan led a successful surprise attack on a Philistine camp, an earthquake appeared to have contributed to the Philistine panic and the Israelite victory: "There was a panic in the camp, in the field, and among all the people; the garrison and even the raiders trembled; the earth quaked; and it became a very great panic" (1 Samuel 14.15). This account may, of course, include some hyperbole in the post-victory euphoria; equally, it may be based on a genuine experience that more than the surprise of a shock attack contributed to the success.

In the New Testament, Matthew's account of Jesus' crucifixion includes an earthquake as one of the signs of this momentous event (Matthew 27. 51, 54), though this is not included in the other gospels.

In another Psalm, an alternative view of earthquake activity is offered. In this case God is seen as a refuge from the dangers of an earthquake rather than its immediate cause:

God is our refuge and strength, a very present help in trouble. Therefore we will not fear, though the earth should change, though the mountains shake in the heart of the sea.
Psalm 46.1-2

The sea represented in the Hebrew mind the residue of the primeval chaos which God had tamed. Canaanite images of the sea and sea monsters in struggle with the creative powers of the land are subsumed in the Israelite traditions into the power of Yahweh over the sea, and combined with the divine conquest of the sea in the exodus from Egypt. The story of Noah's flood also provides a picture of the power of God over the waters, in both using and limiting their powers to destroy. The conquered seas retain some continuing power, though even over the seas and the primeval waters God's will prevails:

> *The voice of the Lord is over the waters; the God of glory thunders, the Lord, over mighty waters ... The Lord sits enthroned over the flood.*
> *Psalm 29.3, 10*

> *You divided the sea by your might; you broke the heads of the dragons in the waters.*
> *Psalm 74.13*

> *You rule the raging of the sea; when its waves rise, you still them.*
> *Psalm 89.9*

> *In his hand are the depths of the earth; the heights of the mountains are his also.*
> *The sea is his, for he made it, and the dry land, which his hands have formed.*
> *Psalm 95.4-5*

> *Whatever the Lord pleases he does, in heaven and on earth, in the seas and all deeps.*
> *Psalm 135.6*

However, a continuing fear of the sea remains, and God is

worshipped because he has limited its extent:

I placed the sand as a boundary for the sea, a perpetual barrier that it cannot pass; though the waves toss, they cannot prevail, though they roar they cannot pass over it.
Jeremiah 5.22

[I] prescribed bounds for it [the sea], and set bars and doors, and said, "Thus far shall you come, and no farther, and here shall your proud waves be stopped".
Job 38.10-11

You set a boundary that they [the waters] may not pass, so that they might not again cover the earth.
Psalm 104.9

These biblical writers expressed confidence in the fixed nature of the land and its resistance against any encroachment by the sea. The coastline was a boundary and a "perpetual barrier" representing God's protection of the land. There was, of course, no perception that on a geological timescale coastlines are continually changing, and represent only the present division between the age-long fluctuations between land and sea.

Although the biblical writers saw a measure of stability and permanence in the earth's foundation, this was based in the will and purposes of God. At some point those purposes could change, indeed would change, and at that time the very physical structure of the earth would be altered. The occurrence of earthquakes was taken as an indication that such changes were within the power of God. As the earth had a beginning in God's creative action, so the earth and the heavens should anticipate an end, or at least a radical transformation:

Long ago you laid the foundation of the earth, and the heavens are

the work of your hands. They will perish, but you endure; they will all wear out like a garment. You change them like clothing, and they pass away; but you are the same, and your years have no end.
Psalm 102.25-27

I am about to create new heavens and a new earth; the former things shall not be remembered.
Isaiah 65.17; 66.22

On that day there shall be a great shaking in the land of Israel ... and the mountains shall be thrown down, and the cliffs shall fall, and every wall shall tumble to the ground. Ezekiel 38.19-20
I will show portents in the heavens and on the earth, blood and fire and columns of smoke. The sun shall be turned to darkness and the moon to blood, before the great and terrible day of the Lord comes.
Joel 2.30-31

On that day, says the Lord God, I will make the sun go down at noon, and darken the earth in broad daylight.
Amos 8.9

the Mount of Olives shall be split in two ... there shall be continuous day (it is known to the Lord), not day and not night ... The whole land shall be turned into a plain.
Zechariah 14.4-10

This expectation of a new order is of course continued in the New Testament, though there is ambiguity as to whether the renewal is of a physical or a non-physical world. In one of the few New Testament passages offering an interpretation of the natural world, the apostle Paul wrote that following the redemption of the "children of God", the creation also will be set free and will enjoy a new freedom (Romans 8.19-22). In this passage, "creation" appears to mean everything in existence other than humans. The destiny of humankind is described separately and is, indeed, the foundation on which the freedom of all other parts of creation is built. The creation has been "subjected to futility" (v20), a reference to the divine curse on the ground in Genesis 3.17-18 – the creation has lost some of the fullness and richness and purpose of the divine intention, and, in Paul's thought, is now governed by inferior powers opposing God's will. As Cranfield comments, "We may think of the whole magnificent theatre of the universe together with all its splendid properties and all the chorus of sub-human life, created to glorify God but unable to do so fully, so long as man, the chief actor in the drama of God's praise, fails to contribute his rational part."[45] Nevertheless, this subjection is "in hope" because the future freedom of the children of God will also result in creation being freed from "its bondage to decay" (vv20-21).

This hopeful view of a glorious and uncorrupted future for the physical and non-human world is difficult to relate to a scientific view of natural processes. Corruption and decay are now understood as necessary to the recycling of living material between generations, not only in the well-known "food chain" but also in the perpetual creation of nutrients in the soil through the decay of vegetable matter and creaturely bodies, which is essential to the continued growth of plants and the production of food crops. Death according to Paul is in opposition to the goodwill of God, but modern biology regards the death of all living things as essential to the continuity of life. The vision of Paul for a future incorruptible world for humans and the non-human creation therefore would entail substantial changes to the operation of scientific principles as presently known.

The resurrection of the body was another essential part of Paul's faith, derived in part from his training as a Pharisee (Acts 23.6; 24.21; 2 Corinthians 5.1-5), though Paul himself struggled to explain the differences between the earthly body and its new resurrected form (1 Corinthians 15.35-57; Philippians 3.21). The analogy with seeds in 1 Corinthians 15 helps Paul's argument, as seeds are transformed from one form into another, but his argument from different forms of flesh (humans, animals, birds, fish) or heavenly bodies (sun, moon, stars) is less pertinent as these do not change from one form into another. The reason for introducing these comparisons seems to be that since there are different forms of existence, why should the idea of a resurrected body be so difficult? This is not a killer argument to a scientific mind, but it does remind us that scientific knowledge has progressively uncovered new and unexpected forms of existence, and may be expected to continue to do so. Paul's underlying belief is that the human body is not an unworthy shell from which the liberated spirit will eventually escape, but an integral part of the human being which is destined for renewal. In common with contemporary Judaism he believed that the present world order was corrupt ("the present evil age", Galatians 1.4) and that it was passing away (1 Corinthians 7.31). The seer of the Revelation saw the sky vanishing like a scroll being rolled up, every mountain and island being removed from its place (Revelation 6.14), and the present world being replaced by a new heaven and earth, "for the first heaven and the first earth had passed away, and the sea was no more" (Revelation 21.1). The sea, often seen in Israel as the last refuge of the forces in opposition to God, is finally eliminated.

In summary, the biblical view of the natural landscape and seascape consists of four main elements. First, God is the maker and controller of all, including, at least on some occasions, those aspects which might appear undesirable from a human perspective, such as earthquakes and storms. These events are interpreted as instruments in divine hands to persuade the people to live in accordance with God's laws. Second, the watery world, including the seas and the waters above and below the earth which were also a part of the Hebrew cosmology, was also under divine control, though here there is an ambiguous tone, with a continuing sense that the waters are an area where some resistance to the divine will remains. Third, the earth is regarded as almost permanent, though the occurrence of earthquakes is a reminder of a disturbing underlying uncertainty – all things are under God's control and subject to his will, though his will is sometimes disturbing for human existence. And fourth, a hope emerges in the Hebrew Scriptures, which is maintained in the New Testament, that God's people may look forward to a new creation, involving the disappearance of all undesirable features of the present world and the emergence of God's perfect rule.

How does this picture compare with a scientific view of the world and its landscape? There is, of course, no awareness in the Bible of the long periods of time during which the earth has been forming. The earth's hills and mountains may appear permanent from the perspective of human lives lasting a few decades, or even of human history occupying a few thousand years, but geological science now informs us that the present landscape is no more than the passing manifestation of a continuing cycle of the movement of tectonic plates, mountain building, erosion and even the shifting of the continents. The writer of the book of Job was aware of the process of erosion, even at the scale of mountains ("the mountain falls and crumbles away, and the rock is removed from its place; the waters wear away the stones; the torrents wash away the soil of the earth", Job 14.18-19), but the earth's timescale of hundreds of millions of years was not appreciated for many centuries after biblical times. The possibility of the emergence and complete disappearance of whole mountain ranges was of course totally unknown. These considerations make the biblical view of the earth and its landscape very restricted.

Moving our attention from geology to cosmology, the present-day scientist of course has no place for the distinction between the "waters under the dome" and the "waters above the dome" (Genesis 1.7). The traditional translation for "dome" has been firmament. The Hebrew word has the sense of a beaten sheet of metal or a hammered out bowl, indicating the solid nature of the sky. The Latin Vulgate translation of the Bible introduced the word *firmamentum* precisely to convey this sense of solidity. Water was perceived to exist both above and below this heavenly firmament.[46] In the modern view, the rain comes from moisture in the atmosphere, springs arise when the water level in saturated aquifers intersects with the surface of the ground, and the rivers and seas play a continuing role in the hydrological cycle and in the erosion and re-formation of land masses. The sea, though it may be awesome to those who have little or no experience of sailing on it, and, indeed, to those who do, is now regarded as an essential and integral part of the natural world which happens to be composed of fluids rather than solids, and therefore operates on somewhat different physical principles. For all that, it is no more and no less a part of God's creation.

Disastrous events such as earthquakes, floods and plagues of locusts present a different type of problem for the modern interpreter (the biblical world seems to have had little experience of other sources of disaster such as volcanoes and tsunamis, though there are some hints, such as the destruction of Sodom and Gomorrah by "sulphur and fire ... out of heaven", Genesis 19.24-28). At one level these events are the result of the physical forces which have shaped the earth for millions of years. The River Jordan rift valley, incorporating the Dead Sea, is the location of the earthquakes described in the Bible, and is part of a major fault system which extends through the Gulf of Aqaba and the Red Sea into the East African lakes. Earthquakes can be expected more frequently here than in many other parts of the world. Similarly, sudden floods and invasions by plagues of locusts are common though unpredictable phenomena in the arid and semi-arid parts of the world such as the Middle East. These facts are difficult to harmonise with the prophetic interpretation of disasters as judgements by God upon a sinful people. The prophetic view may even be seen as opportunistic, using a natural disaster to force home the prophet's conviction about God's response to the waywardness of the nation. (In the same way, some modern preachers will use a natural disaster or an epidemic as evidence of the judgement of God). Such a view is, I suggest, out of sympathy with a scientific understanding of these events. The nature of the landscape, the seascape and their development is one area in which we must conclude that the biblical view is a long way from the modern scientific understanding. Biblical precedent for a judgemental interpretation is certainly available, but it raises for us enormous moral problems about the nature of God, and is best avoided. We will return to this topic in the section "God as the cause of illness and destruction" in Chapter 4.

God, animals and birds

The Bible sees the animal world as an integral part of God's creation and God's control of the whole of existence. God created everything, including all the animals of the land, the air and the sea (Genesis 1.20-25), the fearful sea monsters, wild animals, birds and creeping things (Psalm 148.7, 10). "Every wild animal of the forest is mine" says another Psalm, and also the cattle on a thousand hills and all the birds of the air (Psalm 50.10-11); the context here is not so much a divine love of animals, but a reminder to the people that God does not need their sacrifices, as he owns all the world's animals including those regarded as ritually unclean. Domestic animals may be under some form of human ownership, but all wild animals belong to God, and he does not need to eat their flesh or drink their blood. Therefore the worship that God requires is not sacrifices but thanksgiving and commitment to God's ways (Psalm 50.13-14). However, God is concerned for animals as well as humans: in a Psalm praising God for his covenant faithfulness we read that "You save humans and animals alike" (Psalm 36.6). The story of Jonah even ends on the note that many animals were saved from destruction through God's forgiveness towards Nineveh (Jonah 4.11 – the animals had also participated in repentant fasting and the wearing of sackcloth, 3.7-8. The participation of domestic animals in mourning is known in Persia, though not otherwise in Assyria.[47]) In one of Amos's visions, God was even "forming" the locusts which threatened a devastating plague, though on that occasion God relented (Amos 7.1-3); on another occasion, a locust plague is described as the day of the Lord, with God at the head of his devouring army (Joel chapters 1-2, especially 2.11).

The book of Job, in its crushing revelations to its eponymous hero of the mysteries of the natural world, attributes to God the provision of food for wild birds, even the ritually unclean raven (Job 38.41). Among the strange sights of nature, not understood but clearly observed by the author, are the habits of wild goats and asses, the independence of the wild ox, the uncaring behaviour of the ostrich towards its young, the bravery of the horse in battle and the hunting skills of birds of prey (Job 39). In Psalm 104, God is described as the immediate source of the springs and grass that provide for both wild and domestic animals, and of the trees in which the birds sing and construct their nests: "You make springs gush forth in the valleys; they flow between the hills, giving drink to every wild animal ... From your lofty abode you water the mountains ... You cause the grass to grow for the cattle" (Psalm 104.10-14).

The biblical writers make use of their observations of birds to provide numerous images for human experiences. The Psalmist longs to possess the wings of the dove so that he might escape his worries and fears and fly away to a place of rest (Psalm 55.6), while the night-time watchfulness of the owl is an image of the distressed person who cannot sleep (Psalm 102.6-7). More positively, the eagle's apparently easy flight provides a picture of renewed strength (Psalm 103.5; Isaiah 40.31; cf. Exodus 19.4), though "the way of the eagle in the sky" is one of the things which are too wonderful for the writer of the Proverbs to understand (Proverbs 30.18-19). In several places the protection of God is described as shelter under his wings[48] – a picture of the motherly hen sheltering young chicks, while also bringing to mind the angelic wings of the covenant throne in the Temple. The simile of a sheltering hen was also used by Jesus in his lament for Jerusalem (Matthew 23.37; Luke 13.34).

The consistency in the times of bird migration was observed by the prophet Jeremiah, who contrasts this reliable behaviour with the deceitfulness and unreliability of the people (Jeremiah 8.4-7). Alternatively, the disappearance of the birds and animals was a sign to the prophets Hosea, Jeremiah and Zephaniah of the devastation of the land – Hosea and Zephaniah add that even the fish of the sea are failing (Hosea 4.3; Jeremiah 4.25-26; 9.10; 12.4; Zephaniah 1.3). On the other hand, for Isaiah it is the *presence* of unshepherded flocks and wild animals that reveals the desolation of the city – the creatures mentioned in this connection include sheep, lambs and cattle among the ruins, "howling creatures", ostriches, "goat-demons", hyenas and jackals, the hedgehog, wild asses, hawks, owls, ravens, buzzards and wildcats.[49] On a more positive note, the hovering of birds overhead is a symbol of God's protection (Isaiah 31.5). A raven and a dove played important roles in the story of the escape of Noah and his family from the devastating flood (Genesis 8.6-12), and ravens fed the prophet Elijah during the drought (1 Kings 17.6).

In the New Testament, Jesus repeats the view that God feeds the birds and clothes the grass (Matthew 6.26-30; Luke 12.24-28), and that not a sparrow falls to the ground without God's knowledge (Matthew 10.29; Luke 12.6). The fact that God knows the fate of the sparrows but does not prevent it is perhaps a significant insight into Jesus' view of happenings in the natural world. Doves are both an image of the Spirit descending upon Jesus (Mark 1.10) and a picture of innocence (Matthew 10.16).

The apostle Paul, however, seems to find the Hebrew laws about the humane treatment of animals of value only as metaphors for God's will for the church. For Paul the law permitting animals to eat some of the grain they were threshing (Deuteronomy 25.4) had meaning only in relation to the material and financial maintenance of the church's leaders: "Is it for oxen that God is concerned? Or does he not speak entirely for our sake? It was indeed written for our sake" (1 Corinthians 9.9-10). Even the allegorical interpretations of the rabbis allowed that God had some concern for the animals, along with a greater concern for humans[50]; but here Paul appears to allow no such concession – the text speaks "entirely" for our sake. Fee's argument that Paul is overwhelmed by the eschatological nature of the times and focuses on the human meaning of the text without denying its reference to animals[51] is unconvincing.

God and the weather

The weather affects everyone, and many people have their own mini-theories about weather systems. Some of these theories may be well founded upon years of experience of the micro-climate in specific places. Although these theories may be described as amateur, many people nowadays are also reasonably well informed about the basic principles of meteorology. Weather maps on television and in the newspapers sometimes display isobars and pressure systems, enabling the informed public to form a better understanding of the way the weather will develop. We understand that our weather depends not only on the season of the year but on atmospheric conditions affecting the direction of the wind and the movement of air masses.

In the lands and times in which the biblical story occurred, most of this type of information was not available. The weather was generally more stable and predictable than in areas such as Britain, but even so there was sufficient variety and uncertainty for the weather to be a cause of concern. Annual and seasonal variations were of course well known, and the expected times of rainfall were critical in determining agricultural practice. The general role of the direction of the wind, whether from the sea in the west or the desert in the east and south, was appreciated, but the detailed mechanisms causing variations in the weather were not known. Also, synoptic information, describing the weather systems prevailing over a large area at any given point of time thus assisting prediction, was, of course, not available. Specific events of rainfall or drought in particular locations must have appeared even more arbitrary to the people of biblical times than they do to us.

In the semi-arid climate of the biblical lands, the occurrence of rainfall was critical in a way which is difficult for those who live in well-watered regions to appreciate. The balance between rainfall, infiltration and evaporation in any season would determine the quantity and quality of a season's crop, or whether there would be any crop at all. In the absence of both long-term storage facilities and any serious prospect of imports from elsewhere, rainfall was the vital factor determining plenty or famine.

In these circumstances, faith in a God who had made everything and could control its disposition would have particular application to the occurrence of rainfall. Jeremiah chapter 14 gives a poignant and moving picture of the impact of a drought, and the prophet's interpretation of the event. A mood of gloom and idleness settles over the country and the city: "Judah mourns and her gates languish; they lie in gloom on the ground, and the cry of Jerusalem goes up" (v2). The water cisterns are dry, and even the nobles with servants to carry their water and money to buy it at a high price if possible, are unable to find any (v3). Farmers are in despair, and the wild animals abandon their young (vv4-6). Jeremiah's view is that God is abandoning the people because of their sins. Prayer and fasting will not produce rain (vv11-12) – God requires repentance (v20). In a significant phrase, Jeremiah says that neither idols nor simply "the heavens" can bring the necessary rain (v22). The reference to "the heavens" as unable to give the necessary showers seems to point here to the theoretical possibility of physical processes operating without God's assistance – perhaps through the activity of other gods, or even through "natural" events. Jeremiah appears to reject any such reliance upon any other means of securing the necessary rain. It is God alone who controls events. "We set our hope on you, for it is you who do all this" (v22).

The direct role of God in sending rain is expressed in other places. The celebrated victory of Deborah and Barak against Sisera seems to have depended on exceptional meteorological conditions. The story of the battle is told in Judges 4 in rather prosaic terms, including Jael's slaughter of Sisera with a tent-peg. But the celebration of the victory in the song of Judges 5 points to a more complex set of factors, including a deluge of rain which is seen as being controlled by the stars:

> *Lord, when you went out from Seir, when you marched from the region of Edom, the earth trembled, and the heavens poured, the clouds indeed poured water.*
> *The mountains quaked before the Lord, the One of Sinai, before the Lord, the God of Israel ...*
> *The stars fought from heaven, from their courses they fought against Sisera.*
> *The torrent Kishon swept them away, the onrushing torrent, the torrent Kishon.*
> *March on, my soul, with might!*
> *Judges 5.4-5, 20-21*

The importance of rain in the life of the nation, and the association of drought with sinfulness and the need for repentance, is also shown in Solomon's prayer at the dedication of the Jerusalem temple. "When heaven is shut up and there is no rain because they have sinned against you", and when they pray and turn away from their sins, then forgive their sins, teach them the good way to live, "and grant rain on your land, which you have given to your people as an inheritance" (1 Kings 8.35-36). This repeating cycle of sin, disaster, repentance and renewed blessing is a typical feature of the "Deuteronomist" understanding of history. Most commentators consider that the historical books were authored, or at least edited, from a Deuteronomist perspective. Richard Nelson writes of 1 Kings 8, "This chapter is virtually a primer of Deuteronomistic theology, the dominant ideological perspective in the Book of Kings."[52] The Chronicler's version of this event adds that when God appeared to Solomon after the dedication ceremonies, God himself endorsed this understanding of events: "When I shut up the heavens so that there is no rain ..." then if the people pray and turn from their wicked ways God will forgive and heal the land (2 Chronicles 7.13-14).

The prophet Elijah, in his conflict with King Ahab, claimed that by his prophetic word he was able to maintain a drought for three years: "As the Lord God of Israel lives, before whom I stand, there shall be neither dew nor rain these years, except by my word" (1 Kings 17.1). This claim that Yahweh will determine the occurrence of rain is, of course, a direct challenge to the supposed rain-bringing powers of the rival god Baal. During the drought Elijah was reportedly preserved by water supplies from the Wadi Cherith, by food brought to him by ravens, and by meal and oil miraculously maintained for the widow of Zarephath (vv3-16). The drought was spectacularly ended by Elijah's contest with the prophets of Baal on Mount Carmel, after which Elijah prayed and told his servant to "look towards the sea". Gradually the clear sky produced a cloud no bigger than a hand, and then "the heavens grew black with clouds and wind; there was heavy rain" (1 Kings 18.42-45). The physical association of rainfall with wind from the direction of the sea and with thick clouds is recognised, but the reader is left in no doubt that God is the source of the rain just as God was the source of the fire that consumed Elijah's sacrifice. The power of Elijah's word in withholding and then releasing rain is endorsed in the letter of James in the New Testament – here the prophetic word is understood as prayer: Elijah "prayed fervently that it might not rain, and for three years and six months it did not rain on the earth. Then he prayed again, and the heavens gave rain and the earth yielded its harvest" (James 5.16-18). According to James the effectiveness of the prayer rested not only in the prayer itself but also in the righteous character of the one who prayed.

The association of rain with strong wind, and also with the action of God, is also present in Ezekiel's image of the whitewashed or plastered wall: "Thus says the Lord God: In my wrath I will make a stormy wind break out, and in my anger there shall be a deluge of rain, and hailstones in wrath to destroy it [the wall]" (Ezekiel 13. 11-13). As the prophecy continues, it is clear that the wall is itself an analogy for the walled city of Jerusalem (vv15-16). Again, God is the direct source of rain and hail against the invader Gog: "With pestilence and bloodshed I will enter into judgment with him; and I will pour down torrential rains and hailstones, fire and sulphur, upon him and his troops and the many peoples that are with him" (Ezekiel 38.22).

Other similar examples, attributing either the provision or the absence of rain to God's direct action are:

> *Is it not the wheat harvest today? I will call upon the Lord, that he may send thunder and rain ... So Samuel called upon the Lord, and the Lord sent thunder and rain that day.*
> *1 Samuel 12.17-18*

This thunderstorm was particularly noteworthy as rain is rare at the time of the wheat harvest, thus underlining the direct responsibility of Yahweh, in response to Samuel's prayer, for this event.

The Lord our God, who gives the rain in its season, the autumn rain and the spring rain, and keeps for us the weeks appointed for harvest.
Jeremiah 5.24

O children of Zion, be glad and rejoice in the Lord your God; for he has given the early rain for your vindication, he has poured down for you abundant rain, the early and the later rain, as before.
Joel 2.23

I also withheld the rain from you when there were still three months to the harvest; I would send rain on one city, and send no rain on another city; one field would be rained upon, and the field on which it did not rain withered; so two or three towns wandered to one town to drink water, and were not satisfied; yet you did not return to me, says the Lord.
Amos 4.7-8

The heavens above you have withheld the dew, and the earth has withheld its produce. And I have called for a drought on the land and the hills, on the grain, the new wine, the oil, on what the soil produces, on human beings and animals, and on all their labours.
Haggai 1.10-11

The skies shall give their dew; and I will cause the remnant of this people to possess all these things.
Zechariah 8.12

Ask rain from the Lord in the season of the spring rain, from the Lord who makes the storm-clouds, who gives showers of rain to you, the vegetation in the field to everyone. Zechariah 10.1

Note that Zechariah here commends prayer for rain in the season when it is regularly expected. Does this signify that nothing should be taken for granted, and that God is the author even of those phenomena which appear to be physically regular? Is there a corollary that it would be pointless to pray for rain in the dry season? The examples quoted above show that this was not necessarily the case.

> *If any of the families of the earth do not go up to Jerusalem to worship the King, the Lord of hosts, there will be no rain upon them.*
> *Zechariah 14.17*

Jesus also teaches the direct responsibility of God for the rain:

> *[Your Father in heaven] makes his sun rise on the evil and on the good, and sends rain on the righteous and on the unrighteous.*
> *Matthew 5.45*

The point of this saying appears to be that God is generous and non-discriminatory in his provision of natural resources such as sunshine and rain. It should not be taken to mean that God does not care about the presence or absence of good or evil. These words of Jesus seem to challenge the Deuteronomist and prophetic belief that the occurrence of rain is related to the righteousness of the people.

As we have already seen in some of the above quotations, rain is not the only meteorological phenomenon for which God was regarded as responsible. He is also credited with hailstones ("the Lord threw down huge stones from heaven ... and they died; there were more who died because of the hailstones than the Israelites killed with the sword" Joshua 10.11[53]), thunder (1 Samuel 2.10[54]), and wind (Psalm 78.26; Jonah 1.4). Most controversially of all, on one occasion according to the book of Joshua (10.12-14) God is said to have stopped the sun and the moon in order to extend daylight, enabling Israel to complete a defeat of the Amorites. These verses were hugely important in the seventeenth century debates between Galileo and his ecclesiastical accusers concerning the true motion of the sun and planets, and they still present difficulties for a too-literal understanding of the text. This example of God's control over natural events will be discussed further in the discussion of miracles Chapter 5.

In a New Testament example of authority over weather, Jesus is portrayed as calming or "rebuking" a storm on the lake (Matthew 8.23-27). The background to this story may be seen in the Hebrew accounts of God's dominion over the ancient power of the watery world. In Matthew's account Jesus, having shown his authority over matters of the law ("you have heard that it was said ... but I say to you", 5.21 and following), now proceeds to demonstrate his authority over creation. R. T. France gives a number of parallels between this story and certain Psalms and the story of Jonah, suggesting that Jesus is in the place held by God in the Hebrew texts.[55]

The story of Noah and the universal flood (Genesis 6-9), which I have argued belongs to the genre of folk stories and therefore should not be regarded as historically descriptive, does nevertheless inform us of one way in which the natural world and God's relationship to it were understood. We are told that the flood was the expression of God's intention "to destroy from under heaven all flesh in which is the breath of life; everything that is on earth shall die" (6.17). The flood occurred when "all the fountains of the great deep burst forth, and the windows of heaven were opened" (7.11) – the waters above and below the firmament, which God had separated at creation (Genesis 1.7) were thus brought together again, though the flood is also attributed to forty days and nights of continuous rain (7.12) – a combination of ancient stories and their texts is seen by many commentators as the reason for such discrepancies. The reversal of the process is also ambiguously described: "God made a wind blow over the earth" (8.1), but also the fountains of the deep and the windows of heaven were closed (8.2). In this highly meteorologically oriented story, we see a combination of naturalistic (as understood by the Hebrew author) and divine causes.

Do these many attributions to God of the responsibility for specific natural events amount to a case for divine direction of *all* such events? Is it possible to produce a general philosophy of divine action from these incidents? The prophets' readiness to call upon the Lord to send rain or withhold rain seems to be possible only on the basis that God is the one who controls all these things at all times. And Jesus' statement that God sends sunshine and rain on the righteous and unrighteous alike (Matthew 5.45) appears to be independent of specific instances. In the Psalms and Proverbs we come closer to what may be regarded as a considered view, capable of reference to specific instances but not limited to single events.

> *Whatever the Lord pleases he does, in heaven and on earth, in the seas and in all deeps. He it is who makes the clouds rise at the end of the earth; he makes lightnings for the rain and brings out the wind from his storehouses.*
> *Psalm 135.6-7*

> *He covers the heavens with clouds, prepares rain for the earth, makes grass grow on the hills.*
> *Psalm 147.8*

> *The Lord by wisdom founded the earth; by understanding he established the heavens; by his knowledge the deeps broke open, and the clouds drop down the dew.*
> *Proverbs 3.19-20*

The last of these three texts reveals an interesting development in Hebrew thought. The author of the proverb agrees with the psalmists that the Lord did all these wonderful things, but the proverb adds significantly that the Lord did it by "wisdom", "by understanding" and "by knowledge". This introduction of God's wisdom, understanding and knowledge serves to bring the awesome activity of God just a little closer to human appreciation. The word translated "understanding" (Hebrew *t^{e}bunah*) can mean skill in manual craftsmanship.[56] It remains the case that the deeds of the Lord are beyond human capability and repetition, but the suggestion that God did it through wisdom, understanding and knowledge begins to render his actions just a little bit less arbitrary and more comprehensible. Human beings, after all, may claim to possess a small measure of wisdom, understanding and knowledge, so the inclusion of these terms in the text of Proverbs encourages the reader to feel some affinity with God's actions. God is not totally unknowable or his deeds totally inscrutable. Human self-appreciation is also raised in the process.[57]

God and harvest

A natural extension of the Hebrew view that God controlled the weather was the attribution to God of harvests both good and bad. Laws concerning the celebration of harvest are found at Exodus 23.16-19 and 34.22, and a fuller harvest liturgy is given at Deuteronomy 26.1-11. The latter passage is more concerned to repeat Israel's longer-term history than to present a theology of divine providence: the nation has developed from a period of nomadic wandering, through captivity in Egypt from which "the Lord brought us out", to their present life in the land flowing with milk and honey. The worshipper brings to the priest "the first fruit of the ground that you, O Lord, have given me" (v10). The gift of God appears in this case to be the land rather than the produce – this is explicitly stated a few verses later ("bless your people and the ground that you have given us", v15).

Elsewhere, however, the relationship between God's provision and the results of harvest is expressed more explicitly. In the book of Joel God says "I am sending you grain, wine and oil, and you will be satisfied; and I will no more make you a mockery among the nations" (Joel 2.19). In the context of Levitical legislation for the sabbatical year, God promises extra abundance in the sixth year in order to tide the people over the seventh year of sabbatical rest: "I will order my blessing for you in the sixth year, so that it will yield a crop for three years. When you sow in the eighth year, you will be eating from the old crop; until the ninth year, when its produce comes in, you shall eat the old" (Leviticus 25. 21-22). Whether the sabbatical year was ever anything more than an ideal of pious lawmakers is not known.

The withholding of harvest was also seen as an action of judgment by God. In response to King David's action in making a census of the people, the prophet Gad offers him a choice between three years of famine, three months of military defeat, or three days of plague (2 Samuel 24.13). David chose the last named and a subsequent short but intense epidemic is reported, but the clear implication is that God was well able to ensure three years of calamitous agriculture. Jeremiah says that because of Israel's misconduct, both moral and religious, "the showers have been withheld, and the spring rain has not come" (Jeremiah 3.3), whilst in Moab, "Gladness and joy have been taken away from the fruitful land of Moab; I have stopped the wine from the winepresses; no one treads them with shouts of joy; the shouting is not the shout of joy" (Jeremiah 48.33).

Although, as we have seen, the New Testament is more concerned with human salvation and righteous living than with the practicalities of living in the land which were of more concern to the ancient Hebrews, there is a passing reference by the apostle Paul to "He who supplies seed to the sower and bread for food" (2 Corinthians 9.10; cf. Isaiah 55.10). Significantly, this idea of divine generosity is used by Paul as a foundation for Christian generosity rather than as part of a developed argument concerning providence in nature.

The biblical view and the scientific view

How should the scientifically minded reader react to this biblical view of God's direct control of the natural world, including the weather? Two points should be made at this stage. First, the natural phenomena described (rain, drought, hail, etc.) are now recognised as the expressions of long-term and continuous processes. The occurrence of rain or any other of these events is the result of antecedent conditions stretching back over a long period of time – in fact, if each and every event is the result of its antecedents, the natural world consists of a near-infinite series of causes and effects. This is not to say that the whole natural process is entirely deterministic – there is space for random developments and unpredictability, but the introduction of such factors remains at the natural level. This long and continuing physical process renders questionable the biblical assertion that a particular occurrence of rain or hail or drought is a divine response to particular local conditions – especially to the moral state of a nation, which would appear to have no effect on phenomena such as rainfall. [58]

Second, we have seen that even the Bible texts recognise the role of natural causes of these events. Rain or drought is not caused solely by the arbitrary decision of God – these phenomena are related to the occurrence of particular winds, and the presence or absence of clouds. A physical cause is recognised, but God is seen as active behind (or within?) that physical cause. One is left with the feeling that if more had been known about the physical causes of the weather, the action of God would also have been described differently. In those circumstances, the divine action might have been regarded as further back in the process (the text might have read, "God caused a set of pressure systems, which set up a complex wind circulation, which led to the interaction of air masses of differing temperatures, which caused a great storm and heavy rain ...") or the divine action might have been detached altogether from the physical events, i.e. God might have achieved his will in very different ways.

As noted above in the case of Jeremiah 14.22, there may even be a recognition in the Bible that non-divine forces which Jeremiah there calls "the heavens" may in principle be responsible for natural phenomena. For Jeremiah, these non-divine forces are as unworthy of human trust as "the idols of the nations"; like the idols with which they are in poetic parallelism, these forces may or may not even exist – the important point is to trust only in the Lord. We might refer to these forces as "nature", and we know that natural forces do exist and that they have a profound influence on events – so the substance of our trust in the Lord has to be rather different from the experience of Jeremiah.[59]

At 5.24, Jeremiah tells us that the Lord gives the rain "in its season", which for the region of Israel means generally in spring and autumn. What is the association between the gift of God and rain which occurs in its regular season? This may sound to us like a serious case of double explanation, thanking God for what would probably have occurred "anyway". But such a rational approach makes use of knowledge that was not available to Jeremiah. In his time, the regularity of the rain in spring and autumn had been noted, but the reasons for it were not known. We are aware that these reasons relate to the angle of inclination of the earth's axis to the plane of the earth's orbit around the sun, the consequent seasonal variation of the altitude of the sun and the accompanying latitudinal shift of climatic belts, and the resulting pattern of winds around the eastern Mediterranean. The causative interactions between these factors were not known in Jeremiah's time. It was therefore not unreasonable to rely upon God for the continuation of events which appeared to have a seasonal pattern. For us, however, the relationship between event and faith must be described differently.

The meteorologist Sir John Houghton, formerly Chief

Executive of the UK's Meteorological Office, believes that it is "entirely appropriate to pray about the weather, as it is about anything else that is of concern to us."[60] Houghton's worldview is composed of a "scientific story" which describes the way the world works and a "faith story" which relates to religious beliefs, worship and prayer. In his view, the two stories are complementary: "the existence of the 'scientific story' in no way invalidates the 'faith story'." In discussing possible limitations to prayer, Houghton notes that Jesus said that prayer could move mountains (he does not discuss whether this might be typical Jewish hyperbole) and that with God all things are possible. However, he also notes that Jesus "did not interfere capriciously with the natural order" and that "miracles in the sense of events outside the normal scientific order are surprisingly rare in his life."[61] This approach seems to me confusing. Houghton wishes to retain a "scientific story" and a "faith story" but the relationship between them is not at all clear. Vacillating between one and the other is not a coherent position. If the scientific story and the faith story have to be complementary, would it be acceptable to pray for a particular type of weather in those climatic regions where the weather often changes rapidly, but not in regions with long periods of stable conditions?

Does God control the weather, or not? The Hebrew writers, not having access to the modern science of meteorology, would have said yes. Houghton seems to be saying that normally the weather is determined by scientific principles, but occasionally, just occasionally, God may still a storm or send rain or sunshine. Rather than showing the complementarity of the scientific and the faith stories, this view actually undermines the consistency of the scientific story, renders God arbitrary, and makes faith in God less secure.

4

God and the human world

In many respects human beings are part of the natural world. Thoroughgoing materialists would say that in *every* respect human beings are a natural phenomenon, but that is not the view proposed in this book. Our human bodies have similar though not identical features to those of closely related species. Our biological functions operate in a manner similar to theirs, and are similarly liable to physical damage, disease and eventually death. Human life depends upon phenomena such as eyesight, breathing, eating and drinking, digestion, exercise, sexual reproduction, and other physical functions. Other animal species rely upon the same phenomena, and although the details of the processes and the biochemistry involved may vary considerably, these similarities point to the relatedness of humans to the rest of the natural, animal world. In recent decades, this relatedness has been shown even more powerfully through the discovery of large areas of common patterns in the DNA of humans and other species. These are compelling reasons to accept the prevailing scientific view that humans have emerged and evolved from preceding species.

In other respects, humans are very different from all other species. In the few tens of thousands of years of human history (or the few millions of years if the boundary between our precursor hominids and their ancestor species is drawn somewhat more remotely), mankind has developed speech, the structure of language, technology and philosophy, and our utilisation of the world's resources has expanded vastly more than that of any other species. While the life style of most species remained unchanged for millions of years, mankind radically changes his habitat and life patterns over a period of a few decades or centuries. Moreover, mankind is the only species, as far as we know, that has any sense of God or any belief in realities beyond the immediate physical world, though it is possible that some animals possess a sense of the existence of "unknowns" beyond their experience, similar in some ways to mankind's sense of a spiritual reality.

Our examination of the biblical view of the natural world must include the human, for two reasons. First, as already noted, humans are part of the natural world. Humans are rooted in the physical existence of their bodies, which are themselves immersed in physical processes. Mankind's ability to move beyond his solely physical existence to become aware of a different, spiritual, dimension to his life does not eliminate the reality of his physical nature. Second, and more important, the primary interest of the writers of the Hebrew and Christian Scriptures was in mankind and mankind's relationship to God. The physical world, particularly the land of Israel, was especially important in the Hebrew Scriptures as the location in which God's purpose for his people was worked out, but the main focus of the Scriptures is upon human ethical and religious behaviour. In the New Testament, the teachings of Jesus draw copiously on common events in everyday human life and on observations of non-human nature, but even here the emphasis is on aspects of human or divine behaviour, not on nature for its own sake. And in the rest of the New Testament, interest in the non-human world is almost totally lacking. As far as the New Testament is concerned there are only a very small number of passages from which a "theology of nature" can be constructed.

In this chapter, therefore, I will examine several aspects of human life in which our present scientific knowledge gives us a different viewpoint from the viewpoints of the biblical writers. These aspects will be human nature, gender, the role of dreams, and the perceived role of God as the cause of events both rare and common, desirable and undesirable. The chapter will then go on to examine the interpretation of political events and history, and the respective roles of God and humans in the outworking of these events.

Human nature

"What are human beings?" asked the psalmist (Psalm 8.4). The answer given in the biblical texts is that mankind is a physically based species living in communities, with a rich inner life and striving for a relationship with a higher being named God. For most of the Hebrew Scriptures this relationship with God is expressed partly at the individual level, but mainly through participation in the religious and national life of the community of Israel. Any expectation of a personal immortality is only weakly expressed in a small number of verses. In New Testament times this debate continued: both Jesus and Paul sided with those who believed in a resurrection, but other influential groups maintained the older Hebrew view denying this possibility.

Scientific discoveries about human biology and physiology have of course enhanced enormously our understanding of the workings of the human body and brain, and medical treatments and life expectancy have improved accordingly. However, scientific knowledge is limited to the operations of the body and, to a lesser extent, the psychology of the personality. Scientific enquiry has not found a "soul" located somewhere within the human body, but otherwise science is unable to advise on whether or not the world of the divine, of the spiritual, is a reality. These questions are beyond the range of scientific investigation.

Gender

One area in which the practices and views incorporated in the Bible text differ from our present-day assumptions is in the matter of the respective roles and places in society of male and female persons.

The Priestly creation story in Genesis 1 treats the male and female with equal respect. Here God is said to have created male and female together, and God blesses them and gives them a common purpose (Genesis 1.26-31). The command to the humans to be fruitful and multiply, clearly expressing their sexual differentiation, echoes similar instructions concerning plants, sea creatures and birds (vv11-12, 22). Humanity's intimate association with a fruitful reproducing world is apparent. In the second creation story, however, which is generally regarded as an earlier compilation, a very different picture emerges. The first human is evidently male, and God then creates animals as possible companions for him. The animals proving inadequate for the task, the woman is formed from one of the man's ribs (Genesis 2.7-24) – the story seems to be saying that the woman is superior to the animals, and the man expresses his appreciation of her (2.23), but this remains a long way from the higher view of women in Genesis 1, and from our contemporary view of the equality of both genders. As is frequently noted, the story of the entry of sin in the Garden of Eden in Genesis 3 places the primary responsibility for the initial temptation on the weakness of the woman, and as a result the man is given permanent command over her (3.16[62]). Moreover, the Genesis 2-3 story lacks the appreciation of Genesis 1 that sexuality is a feature that humankind shares with the world of plants and animals – in fact, plants and animals preceded the human species in acquiring sexuality, by hundreds of millions of years in our scientific understanding, though by only a couple of days in the literal sense of Genesis 1.

Although the creation stories present these differing views of the relationship between the male and the female, it was the inferior place of women that dominated in the life of Israel and in the Hebrew Scriptures.

In the Egyptian pogrom against the Israelites, only the baby boys were to be killed – girls were allowed to live, as they could become slaves or wives to the Egyptians and would not convey nationality to their offspring (Exodus 1.16, 22). As Pharaoh's resistance to Moses' demands for freedom began to weaken, he seems to have allowed the Israelite *men* to leave Egypt to worship their God in the desert, for worship was the natural thing for men to do (Exodus 10.7-11) – men are given priority, even in the worship of their gods.[63] These reports of practices in Egypt might be dismissed as non-Israelite, but in fact the same attitudes were present among the Israelites also. Later, as the Israelites purified themselves for the encounter with God at Sinai, the men were ordered not to "go near a woman" – clearly, women were regarded as defiling on the most holy occasions (Exodus 19.15). On the basis of Leviticus 15.16-18, this defilement has been attributed to the male semen[64], but in that case the prohibition of Exodus 19.15 would not be restricted to contacts with the women. Rather, it is the act of sexual intercourse with women that is regarded as defiling, as also in the case of David and his men in 1 Samuel 21.5.

Israelite law contains numerous examples of the inferior status of females. The Ten Commandments includes a neighbour's wife along with his house, field, slaves, donkey and ox as "items" that were not to be coveted – plainly, the Commandments were addressed to men (Exodus 20.17; Deuteronomy 5.21). Only males in the line of Aaron could serve as priests (Exodus 28-29; Leviticus 8-9). Fathers could sell their daughters as slaves, and such female slaves were denied the right of release in the Sabbath year (Exodus 21.7), possibly because it was assumed that they had become the wife or concubine of the master or of a male slave (v4). (The Sabbath year release is, however, extended to both male and female slaves in Deuteronomy 15.12). If a woman gave birth to a male child, she was ceremonially unclean for seven days, but for a female child for fourteen days; similarly the period of "blood purification" was also double for a female child (Leviticus 12.2-5). In a set of obscure laws concerning the value of a human being (perhaps related to the price to be paid to release someone from a vow or from slavery), it is at least clear that males are valued at between one-and-a-half times and two times the value of a female (Leviticus 27.2-8). In this case the difference may be due to the assumption that a man could carry out heavier work than a woman, or would be available for work without interruption due to motherhood, and therefore was of greater value to an employer, though the provisions included infants as young as one month old (v6). In a test case it was ruled that if a man died with no male heirs, his daughters were allowed to inherit (Numbers 27.1-7), but it is noteworthy that a test case was required to establish this principle, and it was also made clear that daughters could inherit only in these particular circumstances (v8) [65]. Although the legal ruling is placed in the context of a particular incident and dated to the time of Moses (the daughters "stood before Moses, Eleazar the priest, the leaders and all the congregation", v2), it is reasonably clear that a dispute of this kind would only occur in a situation of settled landholding. Several of the daughters have names related to the names of Israelite towns[66], which may indicate the locations where the disputes occurred.

However, this ruling led to further complications: if the daughters married husbands from another tribe, their hard-won property would pass to their husbands of the other tribe, thus upsetting the balance of land distribution among the tribes. Thus, in a further ruling (Numbers 36), the same fatherless daughters were allowed to marry whom they chose, but only within their own tribe, in order to maintain ancestral property within the tribe. It is clear that on marriage the women did not retain the property they had previously held.

In another example of male preference, a vow made by a man had to be maintained; women were permitted to make vows, but a man was allowed to overrule a vow made by his wife or unmarried daughter if he did so "at the time that he hears of it", though if he took no immediate action the vow stood (Numbers 30). Budd notes that some of these vows might involve offerings or other commitments, and that the husband or father "must be protected from any excessive commitments made by women who are not ultimately responsible for finding the resources by which those commitments can be honoured."[67] The vows of a widow or divorced woman had the same binding status as those of a man (v9). Gender priority is thus shown to be intimately related to a large accumulation of assumptions about the social roles of men and women.

In the narrative sections of the Hebrew Scriptures the practice of the inferior place of women is again evident. When the inhabitants of Sodom demanded that Lot should release his angelic, presumed male, visitors "so that we may know them", Lot offered instead his virgin daughters, "and do to them as you please; only do nothing to these men, for they have come under the shelter of my roof" (Genesis 19.4-8). As an act of hospitality to visitors this was noble; as an example of gender discrimination it was entirely within the customs of the time. In this case the daughters were spared. In a similar story recounted in Judges 19, the woman concerned dies.

Chapters 9 and 10 of the book of Ezra tell the story of Ezra's demand that the returned exiles should dismiss their foreign wives and their children. This was largely an anti-foreign move, intended to maintain the purity of the nation, rather than anti-women; nevertheless, those who would suffer most from this action were the women and children. It is noteworthy that, according to the account, Ezra spent a night fasting "over the faithlessness of the exiles" rather than over the personal and emotional distress his actions were causing (10.6). Only a few people opposed his instructions (10.15), though the text does not state whether these were more sympathetic to the foreign wives or wished for a more hard-line decision. Williamson opts for the latter[68] though it is difficult to see what more rigorous line could have been taken.

In only a few places in the Hebrew Scriptures is this social bias against women ameliorated by more positive assessments. Proverbs 31.10-31 offers a picture of the "capable wife" who runs an efficient household, trades in the market and buys property, works late on domestic duties, and is generous towards the poor. Even in this instance, the benefit of these worthy actions accrues to her husband, for the quality of the woman is shown in the facts that her husband "will have no lack of gain" (v11) and that "she does him good, and not harm, all the days of her life" (v12). As a result of the good woman's support and careful housekeeping, "Her husband is known in the city gates, taking his seat among the elders of the land" (v23); this was where business was conducted, justice given out, and gossip spread. Because of his stable and prosperous home life, he is able to make a name for himself among the leading citizens. This is decidedly a male view of what constitutes a good woman. On the other hand, the book of Esther is an encouragement to women to be courageous in the face of traditional opposition: the advisers of King Ahasuerus recommend action against Queen Vashti because of her refusal to be paraded as a beauty, and because "this deed of the queen will be made known to all women, causing them to look with contempt on their husbands" (i.e. to refuse their husbands' instructions) (Esther 1.17). The replacement queen, Esther, proves to be equally brave in challenging the convention that no one may approach the king without permission (4.11).

In the New Testament the apostle Paul repeats the traditional cosmological hierarchy of Jewish belief, with Christian amplification: the order of priority is first God, then Christ, then man, and finally woman (1 Corinthians 11.3).[69] This hierarchy is to be demonstrated by decorum in church assemblies, so that a woman must have her head covered when she prophesies (11.5) or, according to 1 Corinthians 14.34-35, must not speak at all. Various attempts have been made to overcome this apparent contradiction: the references may be to different forms of speaking (praying or prophesying is permitted, but chattering during worship is forbidden), or that 14.34-35 are a later insertion by a different writer, or that our present 1 Corinthians is a compendium of several letters written at different times and in different circumstances (an argument that may be supported by considerations other than the gender question). This last solution may mollify the contradiction, but does not eliminate it entirely. The primary sense of these difficult verses seems to be that males reflect the glory of God by leaving their heads uncovered, while women display their modest behaviour and the distinction between the genders by covering their heads. The influence of first century cultural attitudes is clearly very strong here. Whether these customs were Greek or Jewish or both is unclear.

The letters to the Ephesians and the Colossians require women to *be subject to* their husbands, reflecting their legal status under Roman rule, while husbands are to *love* their wives (Ephesians 5.22, 25; Colossians 3.18-19). To be fair, the instruction to love their wives is more considerate than the potential alternative, to rule over them, though that is already implied by the wives' subordinate status. On the other hand, in Galatians 3.28 Paul states that in Christ, i.e. within the shared life of Christian believers, there is no longer Jew or Greek, slave or free, and "there is no longer male and female, for all of you are one in Christ Jesus." While the first two pairs, Jew/Gentile and slave/free, are developed by Paul in a theological way, he is remarkably reticent on the meaning of the ending of the male/female distinction. Moreover, the parallel passage at Colossians 3.11 is in accordance with Galatians in that there is no longer distinction between Greek and Jew, circumcised and uncircumcised, barbarian and Scythian, and slave and free, but does not mention male and female. In the chapter on marriage and celibacy at 1 Corinthians 7, Paul adopts a generally negative attitude towards marriage, though he does at least allow women some decisions in determining their own conduct. In the same manner, Jesus' recorded teachings on divorce, though difficult to interpret with complete consistency, appear to give a greater role to women than was traditionally available.

What conclusions may be drawn from this examination of biblical views about gender, in comparison with our current scientific knowledge? First, any ideas that the female was created later than the male, or that sexuality is a special gift for the human species, must be abandoned. We know that male and female sexuality pervades the plant and animal kingdoms, and that humans are the inheritors of a reproductive system that has been in existence for millions of years. In most species, sexuality is manifested in distinct male and female entities, as with human men and women, but in some species the male and the female exist in other forms (for example, male and female parts of the same entity, or changes over time between male and female gender). Any view that the human female is inferior because of her "later" creation or her proneness to temptation, based on a reading of the Genesis text as though it were literal history, should no longer be accepted.

It is clear that the prevailing biblical view of an inferior place for women was shared with and/or derived from the prevailing views of neighbouring cultures. The ancient origins of this view may be found in the natural restrictions placed upon women during pregnancy, childbirth and childrearing rather than in any divinely ordained distinctions. The inability of women to fight in battles or to undertake arduous physical work at some stages of pregnancy, the significantly increased risk to their lives during labour, and their long-term commitment to childrearing, all served to restrict women's choices until relatively recent times. The views of the biblical writers on these matters were clearly conditioned by the prevailing circumstances of their times. We should recognise that as the biblical view of gender was largely shared with neighbouring cultures, it has less specific relevance to biblical religion.

It is sometimes claimed that the principles of the Jewish and Christian faiths contained, more than any other religious system, the seeds of the emancipation of women. If this is true, it is remarkable that so many centuries elapsed before these seeds germinated, and when they eventually did so it was because of political and sociological rather than religious forces. Even today, a proper recognition of the value and potential of women remains lacking in some Christian contexts, due to over-emphasis on ancient models of male priesthood or culturally conditioned texts in the New Testament. In spite of these evident limitations, the Christian principles that men and women are equally loved by God and equally welcomed in the communion of Christ were vital theological contributions to social change.

Physical, psychological and emotional differences between men and women must still be recognised, though these are no longer expressed in terms of superiority or inferiority. It is also necessary to avoid male and female stereotypes and the accompanying assumptions about standard forms of male and female behaviour and expectations. We are fortunate that the choices now possible in matters such as careers, parenthood and the conduct of family life enable women to live more fulfilled lives than was possible during many centuries.

Dreams

Dreams are now understood as a natural activity of the brain which is essential to mental health. During sleep the brain tidies up earlier experiences, sometimes leading to dream experiences and unlikely combinations of subconsciously recalled events. Deep-seated fears and worries may also be expressed through dreams and nightmares. In earlier times dreams were understood as a means of communication with another world and as messages from God or other supernatural beings. The biblical interpretation of dreams may be explained within the expectations of their times, but the relationship with our present understanding of dreams needs to be examined.

Among the earliest dreams described in the biblical story are those of Abimelech and Jacob. Abimelech, named in the story as King of Gerar, was warned by God in a dream that it was not right for him to have Abraham's wife Sarah in his harem (Genesis 20.3-7). Jacob's dream of a ladder going up to heaven became the foundation of the shrine of God located at Bethel (Genesis 28.11-22). Jacob later attributed his ability to produce speckled goats and sheep, and the order to return to his native land, to the guidance of God in a dream (Genesis 31.10-13), and Jacob's father-in-law Laban also received messages from God in a dream (Genesis 31.24, 29). The provenance of these messages in dreams is intended to show that they had come directly from God, and were therefore indisputable – how different from our present understanding of dreams as processes in the brain derived from earlier thoughts or experiences.

The patriarch Joseph is presented to us as both a dreamer and an interpreter of other people's dreams (Genesis 37-41). The hostility shown towards him by his brothers was due not only to their father's favouritism towards his younger sons, but also to Joseph's boastful recounting of his dreams which portrayed all the brothers and his parents bowing before him. Again, the implication that dreams contained information from God, and therefore that Joseph was receiving personal communication from God, only served to increase the antagonism of the brothers towards him. In the masterly storytelling of the Genesis narrative, this dream was fulfilled years later when Joseph controlled the grain stocks of Egypt and his brothers came to him petitioning for supplies (Genesis 37.5-11; 42.6). Earlier, during his undeserved incarceration, Joseph had been able to interpret correctly the dreams of Pharaoh's butler and baker, which eventually led to the release of Joseph to interpret the dream of Pharaoh himself. As a result, Joseph became the vice-ruler of Egypt, masterminding the management of food during the years of abundance and famine. The old man Jacob again receives guidance through a dream, to join his son Joseph in Egypt (Genesis 46.2-4).

Other positive views of guidance through dreams occur in the stories of Gideon (Judges 7.13-15) and Solomon (1 Kings 3.5-15; 9.2-9 – compare this account with the later retelling of the story in 2 Chronicles 1.7, where the specific reference to a dream is omitted), and in the speech of Elihu to Job, where Elihu regards dreams as one of two ways in which God speaks to humans (Job 33.14-18) – the other way, says Elihu to the suffering Job, is through pain (vv19ff). The promises and warnings of Psalm 89.19-37, where "David", a synonym for any of the later Davidic kings, is assured of God's continuing favour but also threatened with punishment if the people deviate from God's requirements, probably refer back to the dream story of Solomon, here described as a "vision" (v19). The positive role of dreams in revelations to prophets generally is stated in Numbers 12.6.

Towards the close of the period of the Hebrew Scriptures, the figure of Daniel appears as a Joseph-like personality, both a dreamer himself and an interpreter of the dreams of King Nebuchadnezzar (Daniel 2; 4; 7-8).

In spite of these many examples of guidance through dreams, the Hebrew Scriptures do not present us with a uniformly favourable view of dreams and dreaming. The laws of Deuteronomy include prohibitions against prophets making use of dreams to turn the people towards other gods, and against using dreams for purposes of divination even if the predictions turn out to be correct (Deuteronomy 13.1-5). This prohibition is echoed sarcastically by the prophet Jeremiah (23.25-32):

Let the prophet who has a dream tell the dream, but let the one who has my word speak my word faithfully. What has straw in common with wheat? says the Lord.
Jeremiah 23.28

The prophet Zechariah also repeats the condemnation of dreams (10.2), though Zechariah's own visions occurred in dream-like situations (1.8; 4.1). It is difficult to distinguish between the practices forbidden in these negative texts and those regarded favourably in the stories of Joseph and Daniel.

On the other hand, Ecclesiastes offers a typically sardonic view of dreams: they are not particularly useful, but neither are they condemned as totally wrong. Dreams may be more frequent for those burdened with worries; dreams are just one among many phenomena that the Preacher of Ecclesiastes finds empty and meaningless. It is difficult to fit the words of Ecclesiastes about dreams into the neighbouring context of the book – they appear to be isolated sayings:

> *For dreams come with many cares, and a fool's voice with many words ...*
> *With many dreams come vanities and a multitude of words; but fear God.*
> *Ecclesiastes 5.3, 7.*

In the New Testament, dreams are used by God in a positive manner. In Matthew's version of the nativity story dreams are used to inform Joseph that Mary's pregnancy is "from the Holy Spirit" (Matthew 1.20, 24), to warn the magi not to return to Herod (2.12), to instruct Joseph to flee from Herod into Egypt (2.13), and later to return and to settle in Galilee (2.19-22). Matthew also, alone amongst the gospel writers, includes the account of the dream of Pilate's wife and her warning to Pilate to "have nothing to do with that innocent man" (27.19). In Luke's account of the early church, he describes in a positive tone several occasions on which dreams are used to guide the believers: Peter's dream-like vision on the rooftop (at midday "he became hungry and ... fell into a trance" – Acts 10.10) which led to the recognition that Gentiles should be admitted into the body of believers, Paul's vision "during the night" of a Macedonian inviting him to cross the sea into Europe to preach the gospel there (Acts 16.9-10), Paul's vision, again in the night, of the Lord's encouragement at Corinth (Acts 18.9-10), and Paul's account of a trance in the temple at Jerusalem in which he received a vision of Christ (Acts 22.17-21).

This brief summary of the place of dreams in the biblical story reveals no consistent picture or rational explanation. Dreams can be both good and bad (and, presumably, neutral in the great majority of instances). They can be the vehicle for God's guidance, and also for false directions or misinformation. Dreams are woven into the story in a manner which suits the intentions and the unfolding narrative of the respective author. A few major events are accompanied by dreams, such as the birth of Jesus (in Matthew's account, though not in that of Luke, who includes in his account of the nativity what appear to be waking visions rather than dreams), and the spread of Christianity to the Gentiles and into Europe. But for the most part, dreams are not given a significant place in the biblical story. It is evident that some of the biblical authors regarded dreams as a means by which God sometimes guided his servants (parts of Genesis, Daniel, Matthew, Luke in Acts), but in other places the text warns against the potentially harmful influence of dreams.

It may be suggested that it is necessary for us to be sure whether or not God spoke to these biblical characters through dreams, and if so whether God still speaks through dreams. My response to these questions is that our biblical accounts are stories, written by someone other than the characters concerned, and in the case of Genesis written probably centuries after the purported events described. So we cannot be certain about what "actually" happened. The authors of some of the biblical texts appear to have believed that God spoke through the dream stories they were recounting, following the expectations of their times. Now that we have a more scientifically informed understanding of dreams, it remains an open question whether God works in this way.

God as the cause of rare events

It hardly needs to be emphasised that the Bible regards God as the source and cause of rare and unusual events. These events include not only those we might describe as miracles, which will be discussed in more detail in chapter 5, but also events which might be interpreted in a more "natural" way but which have been raised in the telling to a more spectacular level. In several Psalms, the one who can do marvellous things almost becomes a definition of God:

What god is so great as our God? You are the God who works wonders; you have displayed your might among the peoples. With your strong arm you redeemed your people.
Psalm 77.13-15

you are great and do wondrous things; you alone are God.
Psalm 86.10

Remember the wonderful works he has done, his miracles and the judgements he has uttered.
Psalm 105.5

who alone does great wonders.
Psalm 136.4

One generation shall laud your works to another, and shall declare your mighty acts.
On the glorious splendour of your majesty, and on your wondrous works, I will meditate.
The might of your awesome deeds shall be proclaimed.
Psalm 145.4-6

Praise him for his mighty deeds.
Psalm 150.2

And in the later chapters of Isaiah:

When you did awesome deeds that we did not expect,
you came down, the mountains quaked at your presence.
Isaiah 64.3

The awesome deeds were mainly the formative events in Israel's history (see the litany in Psalm 136), but they would also include more personal events such as special births. The birth of Isaac to Sarah and Abraham (Genesis 17-18, 21), and of Samuel to Hannah and Elkanah (1 Samuel 1-2) are treated as special interventions by God even though they were conceived through the normal processes. The marvellous aspect is due to the age of the parents in one case, and the long period of childlessness in the other. These are not miraculous births in the sense of the virginal conception of Jesus described in the gospels of Matthew and Luke; God's action is seen in the fulfilment of his promise of descendants to Abraham, and in answer to the prayerful pleading of Hannah and the prophetic promise of the priest Eli.

The special events associated with the exodus from Egypt may also have "natural" explanations, some of which are hinted at in the text. The plague of boils is attributed to an air-borne pollutant (Exodus 9.8-10), though the origin of "fine dust over all the land of Egypt" from a few handfuls of soot thrown into the air by Moses is more fanciful. Perhaps Moses was claiming by his prophetic action that the unhealthy atmosphere was related to the demands of the God of Israel. The arrival of the hordes of locusts is attributed to a sustained wind from the east (Exodus 10.13). Frequent attempts have been made to show that the succession of the plagues can be related to the natural conditions remaining after each previous event: the pollution of the River

Nile caused the frogs to escape from the river, the decay of the frogs' bodies caused infestations of insects, the insects brought disease to cattle, and so on. If the plagues were due to such natural causes, their special character becomes their cumulative effect on the morale of the Egyptian leaders and their consequence in the freeing of the Hebrew slaves. Against this approach, however, is the widespread critical view that the present text is based on a variety of literary sources containing different plague stories, thus undermining any explanation based on one plague leading by natural means to the next. The retelling of the story of the plagues in Psalms 78.42-51 and 105.26-36 do not include the same destructive items nor follow the same order as each other nor of our present Exodus text.

Similarly, the spectacular escape of the people through the Red Sea (or the Sea of Reeds) is explained by a strong east wind which drove back the sea (Exodus 14.21) – a distinction within this verse between the Priestly text (P) in which Moses the miracle worker is responsible for the division of the waters and the Yahwist text (J) in which Yahweh drives back the sea by means of the strong wind is noted by some commentators. Whatever the literary and historical background to the story, it remains the case that exceptionally low water levels are a well-known hydrological phenomenon in shallow seas affected by strong winds from a particular direction. The poetical description of the waters as a wall on their right and left (vv22, 29; cf. 15.8) gave credence to later images of a more dramatic phenomenon.

Natural causes may have played a role even when not explicitly mentioned. In a retrospective summary of events during his leadership, Joshua says that victory over the Amorites was due to an attack by hornets (Joshua 24.12), even though this is not mentioned in the earlier accounts of the conquest.

The actions of God in moving non-Israelite leaders to perform his will are shown in the cases of Cyrus, the Persian conqueror who released the exiles from their captivity in Babylon, and Artaxerxes who continued the process in the time of Nehemiah. God has taken Cyrus by the hand, says Deutero-Isaiah, and given him stunning victories in his campaigns in order to set the exiles free (Isaiah 44.28-45.13; Ezra 1.1-8).

God as the cause of common events

The recognition of God as the source of exceptional conceptions and births, such as those by Sarah and Hannah, is a specific case of the view that God is the source of all conceptions and childbirths. Not only of Sarah, but also of Leah and Rachel among the matriarchs it is said that "the Lord opened her womb" (Genesis 29.31; 30.22), even though the process was apparently normal. At an earlier stage Jacob attributed Rachel's barrenness to the action of God (Genesis 30.2), and the same is said of Hannah, before the Lord "remembered her" (1 Samuel 1.5-6, 19). Of Ruth also it is said that "the Lord made her conceive" (Ruth 4.13); as Linafelt comments, "Smack in the middle of this cozy romantic scene between Ruth and Boaz one finds the Lord."[70] Several further references to the direct role of God in conception and childbirth are found in the Psalms and prophets.[71] The manner in which any failure to conceive was attributed to the woman is a further example of the biblical attitude towards gender discussed above.

In addition to conception and childbirth, other common events are attributed to God – at least when this suits the theological viewpoint of the writer's narrative. Thus, Samson's attraction towards a Philistine woman was "from the Lord" who was seeking an opportunity to defeat the Philistines, even though Samson's parents did not recognise that their son's emotions were "from the Lord" (Judges 14.1-4). Did God really need "a pretext to act against the Philistines" (v4)? This statement is clearly a later interpretation of the role of God in the events described and, we might say, a somewhat naive one.

In one of the many famines afflicting the region of the biblical events, the Lord had provided food for Israel (Ruth 1.6)[72]. When David was able to remove the spear and water jar from beside Saul's sleeping body, the deep sleep had come from the Lord, even though David did not take advantage of this opportunity to kill Saul (1 Samuel 26.12). Later David "perceived that the Lord had established him king over Israel" (2 Samuel 5.12), even though David himself had not exactly been inactive in working towards this objective. In fact the whole of the story of David's rise to power can be read as an attempted justification of David as selected for the kingship by God and not having pushed for it himself. David's victories are described as a gift from God: "The Lord gave victory to David wherever he went" (2 Samuel 8.6, 14). A generation later, the fame and wisdom of David's son Solomon are attributed to God ("fame due to the name of the Lord", 1 Kings 10.1; "his wisdom, which God had put into his mind", 1 Kings 10.24).

Another example of God's responsibility for a not uncommon occurrence is a theme of the Psalms in which God is regarded as the source of personal or national prosperity:

> *You prepare a table before me in the presence of my enemies; you anoint my head with oil; my cup overflows.*
> *Psalm 23.5*

This Psalm is one of the best known and most popular in the whole Bible, yet it contains a curious and difficult transition from the image of God as the shepherd in vv1-4 to God as host at a generous banquet (vv5-6). The link may lie in the idea of God's provision – the psalmist moves from the metaphor of green pastures for sheep to the reality of sufficient food for humans, or perhaps the banquet is itself a metaphor for the many aspects of the Lord's "goodness and mercy" which will be present throughout his life. The point of immediate interest is that the psalm attributes to God the provision of immediate benefits.

Other psalms do the same:

> *Who are they that fear the Lord? He will teach them the way that they should choose. They will abide in prosperity, and their children shall possess the land.*
> *Psalm 25.12-13*

Psalm 25 is a psalm with an acrostic structure. Fearing or reverencing the Lord will lead to a particular way of living which will result, says the psalmist, in prosperity for both the God-fearer and his following generation. This is an individual expression of the covenant foundation of the nation – if Israel follows the Lord and respects his covenant, the nation will enjoy peace and prosperity.

> *Happy are those who fear the Lord ... Wealth and riches are in their houses.*
> *Psalm 112.1-3*

> *The Lord bless you from Zion. May you see the prosperity of Jerusalem all the days of your life.*
> *Psalm 128.5*

> *May our barns be filled with produce of every kind; may our sheep increase by thousands, by tens of thousands in our fields, and may our cattle be heavy with young.*
> *May there be no breach in the walls, no exile, and no cry of distress in our streets.*
> *Happy are the people to whom such blessings fall; happy are the people whose God is the Lord.*

> *Psalm 144.13-15*
> *Riches and honour come from you, and you rule over all. In your hand are power and might; and it is in your hand to make great and*

to give strength to all ... all things come from you ... all this abundance that we have provided ... comes from your hand and is all your own.
1 Chronicles 29.12-16

All your children shall be taught by the Lord, and great shall be the prosperity of your children.
Isaiah 54.13

Another area in which God is thanked for common events is deliverance from illness or death or the dangers of battle:

You are the one who lifts me up from the gates of death.
Psalm 9.13

The cords of death encompassed me; the torrents of perdition assailed me;
the cords of Sheol entangled me; the snares of death confronted me.
In my distress I called upon the Lord; to my God I cried for help.
From his temple he heard my voice, and my cry to him reached his ears.
Psalm 18.4-6 (cf. Jonah 2.2-7)

Happy are those who consider the poor; the Lord delivers them on the day of trouble.
The Lord protects them and keeps them alive ...
The Lord sustains them on their sickbed; in their illness you heal all their infirmities.
Psalm 41.1-3

The body of Psalm 41 is a declaration that God delivers those who turn to him in their time of distress. But the opening words appear to set a more cautious criterion for God's assistance - God will help those who themselves have acted favourably towards the poor. Deliverance by God cannot be presumed upon, but is the experience of those who have themselves lived in the pattern of divine generosity.

I am poor and needy, and my heart is pierced within me.
I am gone like a shadow at evening; I am shaken off like a locust.
My knees are weak through fasting; my body has become gaunt.
I am an object of scorn to my accusers; when they see me they shake their heads.
Help me, O Lord my God! Save me according to your steadfast love.
Let them know that this is your hand; you, O Lord, have done it.
Psalm 109.22-27

O Lord, my Lord, my strong deliverer, you have covered my head in the day of battle.
Psalm 140.7

According to Acts 26.21-22, in his defence before King Agrippa, the apostle Paul attributes to God his protection from the attacks of the Jews, a rhetoric liable to offend the Roman military who had in fact protected him from Jewish threats, and bound to stir up greater opposition from those Jews who, in trying to end Paul's life and preaching, thought that *they* were God's people doing God's will.

God as the cause of illness and destruction

In my research for this study, I was surprised to discover that the overwhelming majority of biblical examples of divine activity related to the actions of God as the cause of damage, devastation, illness, defeat, and destruction. My notes and references on these aspects of the Scriptures were several times longer than my notes on the activity of God in the natural world, or on God's role in controlling the weather and the harvest, or on divine activity in rare events or common events. The Psalms and the prophets are the major contributors to this material, but the historical books, the teaching of Jesus recorded in the Gospels, the New Testament epistles and the book of Revelation also make their contributions.

This material may be divided into two types. First, there are texts relating to *health and illness*, including the deaths of individuals, where the biblical account is offering a theological explanation of events which could also be understood from a biological or epidemiological perspective. In these cases, the ground for debate lies in the fields of theology and biological science. A second category of material, particularly prominent in the Hebrew Scriptures, describes *historical events*, whether in the past or contemporary with the writer, which are described in terms of divine activity. Here the debate concerns the role of God in the outworking of historical processes – a different question from the relationship of God to natural physical processes. Nevertheless, these two types of texts can both be regarded as examples of potential difficulties when the Bible is read from a modern scientific perspective.

In practice it is often difficult to separate these two types of material. The death of a king from an illness may be regarded as a biological-scientific process, while the death of a king in battle is a piece of human history, but the consequences might be the same as far as subsequent events are concerned.

I will concentrate first on the portrayal of God as the cause of illness and destruction in individual lives, thereby focusing on the interaction of theological interpretation with biological processes. I will then examine larger events affecting greater numbers of people. Also included here is the role of God as the source of destructive "natural" events, such as the Great Flood. Because of the large volume of material, exhaustive quotation is impossible. A few examples will have to suffice.

God's destructive actions towards individuals begin early in the biblical story, with the patriarchs in the book of Genesis. Of each of Judah's two sons, Er and Onan, we are told bluntly that "The Lord put him to death" (Genesis 38.7, 10), the former because he was "wicked in the sight of the Lord" and the latter because he would not fulfil the obligations of levirate marriage for his deceased brother. Later in the Pentateuch it is God who "puts" a leprous disease in a house (Leviticus 14.34), and in Numbers 12 the "anger of the Lord" causes Miriam to become leprous. The direct hand of God in causing individual death is also described in the cases of Abimelech ("Thus God repaid Abimelech ...", Judges 9.56) though the immediate cause of Abimelech's death was a fatal injury from a falling millstone, completed by a sword stroke from his armour bearer; the sons of Eli ("it was the will of the Lord to kill them", 1 Samuel 2.25); Nabal ("the Lord struck Nabal, and he died", 1 Samuel 25.38); Uzzah ("God struck him there because he reached out his hand to the ark; and he died there beside the ark of God", so the place was called Perez-uzzah, the Lord's bursting out upon Uzzah, 2 Samuel 6.7-8); Saul ("the Lord put him to death", 1 Chronicles 10.14, a more direct attribution to God than in the measured account in 1 Samuel 31, although the Chronicler notes in the preceding verse that Saul "died for his unfaithfulness", 10.13); Jeroboam ("the Lord struck him down, and he died", 2 Chronicles 13.20); and the prophet Hananiah ("I am going to send you off the face of the earth. Within this year you will be dead ... In that same year, in the seventh month, the prophet Hananiah died", Jeremiah 28.16-17). In some of these cases the death is immediate, and in most a natural or violent cause of death is given. The writers were clearly interpreting the course of events, whether the progress of a disease or violence in battle, as an expression of the purposes of God, who may therefore be described as the real author of the death. Later theologians might have described this process less personally, by using words such as Divine Providence, but the Hebrew writers were more direct. One commentator describes their interpretation of the deaths of the sons of Eli as "distasteful to modern sensibilities" though "by no means unusual in the Old Testament ... the events of history are directed by Yahweh with specific purposes in view."[73]

More examples are available of God's action in bringing death and destruction on groups, armies or whole nations. Again the story begins early, in the Great Flood. Because of the wickedness of humankind, God says:

> *I will blot out from the earth the human beings I have created – people together with animals and creeping things and birds of the air, for I am sorry that I have made them ... I have determined to make an end of all flesh, for the earth is filled with violence because of them; now I am going to destroy them along with the earth.*
> *Genesis 6.7, 13*

And, according to the story, so God did, with the exception of a single family. As I have explained in chapter 1, I regard the story of Noah's Flood as a folk-tale and not a real description of any historical event. Nevertheless, the story is indicative of Hebrew thought about how God could have acted. A little later in the patriarchal story:

> *The Lord rained on Sodom and Gomorrah sulphur and fire from the Lord out of heaven.*
> *Genesis 19.24*

In Numbers 16 we are told that following the revolt of Korah and others against the leadership of Moses and Aaron, "fire came out from the Lord and consumed the two hundred and fifty men offering the incense" (v35), and in the subsequent "wrath from the Lord" in the form of a plague, 14 700 died (vv46, 49). Deuteronomy 28 sets out the curses which God will bring on the people if they break the terms of his covenant, and we are even told that "the Lord will take delight in bringing you to ruin and destruction" (Deuteronomy 28.63). This may sound harsh and cruel, but the intensity is softened slightly since the use of "take delight" is a deliberate repetition of the preceding sentiment in the same verse, in which God took similar delight in making Israel numerous and prosperous. God "terrified" the Philistine people of Ashdod and "struck them with tumours" when they captured the ark, and God also "killed seventy men" who did not share in the celebration when Beth-Shemesh became the ark's temporary resting place (1 Samuel 5.6; 6.19).[74] Following David's census of the people, "The Lord sent a pestilence on Israel", in which 70 000 died, even though the census was itself incited by "the anger of the Lord" against Israel, presumably because of some unspecified sin (2 Samuel 24.1, 15).[75] The Lord "sent" lions which killed immigrants living in the cities of Samaria (2 Kings 17.25-26), and the Lord also "sent" marauding bands against Judah "to destroy it ... Surely this came upon Judah at the command of the Lord" the writer adds, seemingly to anticipate any contrary view (2 Kings 24.2-3). For the writer of 2 Kings, however, God is "the 'prime mover' in all historical activities."[76] Nelson comments: "The narrator wishes to leave no doubt as to the real cause of Judah's defeat and exile. God did it. The surprising grammatical subject change of 24.2 makes this clear. It was really God and not Nebuchadnezzar who sent this international assortment of raiders to soften up Judah."[77] Even the idyllic story of Ruth, Naomi and Boaz includes the statement that "the hand of the Lord has turned against me ... the Almighty has dealt bitterly with me" (Ruth 1.13, 20-21), though this statement is presented as Naomi's understanding of events and is not necessarily to be taken as the view of the author of the book of Ruth.[78] By the end of the story Naomi's bitterness had been sweetened by the birth of a step-grandchild.

The above examples of God as the source of death and destruction relate to specific incidents. It need not necessarily be inferred that the writers believed that all such events were attributable to God in this way. However, in places, a more generalised view of divine activity of this type is given. In Exodus 4.11, the Lord himself is made to say, when Moses had questioned his own ability to be the Lord's spokesman, "Who gives speech to mortals? Who makes them mute or deaf, seeing or blind? Is it not I, the Lord?" The argument here is from the general to the particular – since God is the one source of deafness and blindness, speech and dumbness, he is able to give Moses the gift of eloquence, which Moses was questioning. Similarly, the Song of Moses contains what appear to be regarded as permanent truths about God, echoed later in the Song of Hannah:

See now that I, even I, am he; there is no god besides me.
I kill and make alive; I wound and I heal;
And no one can deliver from my hand.
Deuteronomy 32.39

The Lord kills and brings to life; he brings down to Sheol and raises up.
The Lord makes poor and makes rich; he brings low, he also exalts.
1 Samuel 2.6-7

In the same vein, Psalms, Lamentations and Deutero-Isaiah offer what appears to be a general philosophy of life and of death:

It is God who executes judgement, putting down one and lifting up another.
Psalm 75.7

Who can command and have it done, if the Lord has not ordained it?
Is it not from the mouth of the Most High that good and bad come?
Lamentations 3.37-38

There is no one besides me; I am the Lord, and there is no other. I

form light and create darkness, I make weal and create woe; I the Lord can do all these things.
Isaiah 45.6-7[79]

As House writes of the Lamentations verse, "The speaker seems to realize that if God is not in control of the bad things (... seen from a human perspective), then he may also not be in control of the good things (seen here as relief from the current situation). In short, the fact that God sent judgment means he can restore their fortunes." [80]

In other texts, a more delicate understanding of the role of God in good and harmful events is apparent. Psalm 30.5 says that "his anger is but for a moment; his favour is for a lifetime"; experience shows that even the greatest tragedy can be faced, and life be renewed afterwards. Immediately preceding the verses in Lamentations quoted above we read that "he does not willingly afflict or grieve anyone" (Lamentations 3.33). In Ezekiel God says, "I have no pleasure in the death of the wicked" (33.11) – God would prefer them to turn from their ways and live. Although Jeremiah had severe messages for God's people, he also records God as saying of the northern tribes:

Is Ephraim my dear son? Is he the child I delight in?
As often as I speak against him, I still remember him.
Therefore I am deeply moved for him; I will surely have mercy on him, says the Lord.
Jeremiah 31.20[81].

By New Testament times, the body of Hebrew Scriptures had come to have authoritative status. Therefore the possibility of divine action to cause death and destruction is accepted, though in many cases the judgements have moved from an earthly to a spiritual plane. Jesus exhorts his disciples not to fear those who can destroy the body, but "rather fear him who can destroy body and soul in hell (Gehenna)" (Matthew 10.28, cf. Luke 12.4-5). Gehenna was the unclean rubbish tip outside Jerusalem that had become a symbol for God's eternal judgement. In his more apocalyptic statements, Jesus warns of eternal fire[82] and weeping and gnashing of teeth[83], and in his parables of useless plants being burned.[84] The background for these sayings may well lie in the permanently smouldering rubbish fires of Gehenna, but even so the prospect is unpleasant. The emphasis does appear to have shifted to the afterlife and away from the here and now, at least in the way in which these sayings were preserved by the church and in the gospels.

In relation to events in this earthly life, Jesus does affirm that the man born blind was not suffering for either his own or his parents' sins, thus leaving open the opportunity for a physical explanation (John 9.3). According to Sanders and Mastin, "Jesus refuses to discuss the cause of the man's affliction". [85] The blindness was an opportunity for God's work to be displayed, as was the case also in the death of Lazarus (John 11.4). Similarly, in commenting on the fate of eighteen persons killed by a falling tower, Jesus says they were not suffering because of their individual sins (Luke 13.4-5). The apostle Paul follows the Jewish methods of interpreting events when he sees in (undisclosed) tragedies the outworking of God's judgment: "God's wrath has overtaken them at last" (1 Thessalonians 2.16).[86] And in Corinth, "many of you are weak and ill, and some have died" because they participated in the Eucharistic meal without discernment (1 Corinthians 11.30). Paul is willing to say that illness and death have occurred because of the abuse of the Lord's table.

Prophecy and the interpretation of political events

Many of the biblical writers display a particular understanding of political events, in the light of their belief in the direct activity of God. Past, present and even future events were seen as the expression of God's purposes in the world, and as God's responses to the faithfulness or unfaithfulness of the people to God's ethical requirements. In this section I will examine the phenomenon of prophecy, and also compare the biblical methods of understanding past and contemporary events with modern practices of interpretation. How do these differences affect our reading of biblical histories?

The basic meaning of the term prophecy is "speaking out" about contemporary events rather than predicting the future. The biblical prophets were primarily concerned about the moral and religious condition of their contemporaries rather than about the distant future or even the near future. Nevertheless, there is a strand of predicting the future in biblical prophecy. The ability to prophesy future events correctly is highly regarded in some biblical books, while prophecy which turns out to be incorrect is regarded as the sign of a "false prophet". Both the backward-looking and forward-looking aspects of prophetic interpretation will be considered in this section.

In the critical theology of the last couple of centuries, prophecies that are too specific have come to be regarded as suspect, i.e. prophecies after the event. Attempts at prophecy are no less common today than previously (for example, political futures, economic trends, climate change) but in most cases they are taken less seriously and failed prophecies are generally forgotten.

The Hebrew Scriptures tend to interpret the events of history, whether in the past or anticipated for the future, as actions by the Lord God. A typical example is found in the treatment in 2 Kings 13.3 of the reign of Jehoahaz, "The anger of the Lord was kindled against Israel" so that they were defeated by their enemies. This is the viewpoint throughout the Deuteronomist history; see also Joshua 7.1; Judges 2.11-21; 3.8; 10.7. In rousing terms, the prophets Hosea, Amos, Micah, Nahum, Habakkuk, Haggai and Malachi also affirmed that God is in action in causing disasters and destruction for Israel. Although not always stated, it was assumed that these punishments occurred because the people had not adhered to the requirements of the covenant. God was regarded as directly and morally responsible for the nation's suffering – note the emphatic occurrence of the first person, that is God, in the following examples:

> *I will punish the house of Jehu for the blood of Jezreel, and I will put an end to the kingdom of the house of Israel.*
> *Hosea 1.4*

The "blood of Jezreel" refers here to the deaths of the kings of both Judah and Israel in a battle near that town instigated by the northern leader Jehu. The reign of Jehu will end, says Hosea, as suddenly and in similar manner as those which Jehu himself ended (even though Jehu's actions against the kings were initially praised by Yahweh, see 2 Kings 9-10, cf. 2 Chronicles 22.7).

> *I will be like a lion to Ephraim, and like a young lion to the house of Judah.*
> *I myself will tear and go away; I will carry off, and no one shall rescue.*
> *Hosea 5.14*

As Stuart states, "The syntax emphasises that God is the lion. 'It is I' and 'I-yes I' begin each half of the verse respectively." [87] God is understood to be directly and violently the cause of the nation's suffering.

> *You only have I known of all the families of the earth; therefore I will punish you for all your iniquities ... I will take you into exile ... I will deliver up the city.*
> *Amos 3.2; 5.27; 6.8*

The Lord commands, and the great house shall be shattered to bits.
Amos 6.11

The Lord has a controversy with his people ... I have begun to strike you down ... Therefore I will make you a desolation.
Micah 6.2, 13, 16

I am against you, says the Lord of hosts.
Nahum 2.13; 3.5

I am rousing the Chaldeans ... The cup in the Lord's right hand will come around to you, and shame will come upon your glory!
Habakkuk 1.6; 2.16

You have looked for much, and, lo, it came to little; and when you brought it home, I blew it away.
Haggai 1.9

And now, O priests ... If you will not listen ... then I will send the curse on you ... I will rebuke your offspring, and spread dung on your faces, the dung of your offerings, and I will put you out of my presence.
Malachi 2.1-3

In all the above examples the prophet ascribes the people's sufferings to the action of God. The "I" of their God Yahweh will also be the author of their destruction.

The prophet Jeremiah has given his name to gloom and pessimism, but a closer look at the book bearing his name reveals that he too sensed that God was the author of the destruction that Jeremiah foresaw coming on Judah and Jerusalem. Jeremiah was not so much a pessimist as a visionary. The very beginning of his book clarifies that it is God who has called the kingdoms of the north to attack the city (1.15). He continues, "Shall I not punish them for these things? [the dishonesty and injustice of the people] says the Lord; and shall I not bring retribution on a nation such as this?" (5.9, 29). The Lord says, "This is the city that must be punished" (6.6). The punishment will include the loss of their homes (6.12), the death of children, parents and old people together (6.11, 21; 13.14), the destruction of the temple (7.14), and the dispersal of the nation (7.15[88]). The means of destruction will include snakes (8.17, possibly metaphorical for enemy forces), fire (11.16; 17.27), drought (14.1-10), sword, famine and pestilence (14.12, 16[89]), dogs, devouring birds and wild animals (15.3), and poisoned water (23.15). God may use these various natural and human forces of destruction, but the prophet leaves no doubt that God is directly responsible for the disasters that befall the people.

In summary, Jeremiah's view of events is placed ironically in the mouth of the captain of the guard responsible for his imprisonment, "The Lord your God threatened this place with this disaster, and now the Lord has brought it about" (40.2-3). Keown et al comment: "A Babylonian soldier was able to identify the sins of Judah as the cause for Jerusalem's destruction ... This is precisely the jarring note that prophets occasionally use to indicate the radical nature of Yahweh's dealings with Israel/Judah." [90] However, lest Israel's opponents should gloat over this devastation, the later chapters of Jeremiah invoke similar wrath from the Lord against Egypt, Philistia, Moab/Edom, the Ammonites, Damascus, Kedar, Hazor, Elam and even the conqueror Babylon itself (chapters 46-51).

The direct responsibility of God for the sufferings of the people is stated even more clearly in the book of Lamentations. According to 1.5, 12-15, "the Lord has made her suffer", the Lord has inflicted, sent fire, handed over, rejected and trodden as in a wine press. "The Lord in his anger has humiliated daughter Zion", destroyed, cut down, killed, become like an enemy, abolished, scorned, ruined and demolished (2.1-9, 17).

> *He is a bear lying in wait for me, a lion in hiding; he led me off my way and tore me to pieces ... He has made my teeth grind on gravel, and made me cower in ashes ... I have forgotten what happiness is.*
> *Lamentations 3.10-11, 16-17*

There is no doubt in the writer's mind about who is responsible for the calamity that has afflicted Jerusalem. As Claus Westermann comments, if it is possible to distinguish between anger as an expression of emotion and as a reaction to events, the destructive actions of God described in the Hebrew Bible are of the latter type. [91]

The prophet Ezekiel, from his viewpoint among the exiled people in Babylon, also attributes to God the sufferings of the people. "I, I myself, am coming against you", says God emphatically. I will do to you what I have never done before, I will execute judgements, I will cut you down, I will have no pity, I will scatter you to the winds, I will vent my fury, I will make you a desolation, I will loose arrows against you, I will bring famine and wild animals and pestilence against you – "I, the Lord, have spoken" (5.8-17[92]). As with Jeremiah, the judgements of God fall not only on Israel but also on the nations of the region (Ezekiel chapters 25-32, 35, 38-39).

The prophet-governor Nehemiah repeats the cyclic view of history which was central to the Deuteronomist pattern of interpretation:

> *They were disobedient and rebelled against you and cast your law behind their backs and killed your prophets ... Therefore you gave them into the hands of their enemies ... Then in the time of their suffering they cried out to you ... you gave them saviours ... But after they had rest, they again did evil before you ... you have dealt faithfully and we have acted wickedly.*
> *Nehemiah 9.26-33*

Some of the Psalms also see God as the real source of death and destruction, upon Israel's enemies and also upon Israel itself if the nation pursues its course of disobedience. For example:

> *If one does not repent, God will whet his sword; he has bent and strung his bow;*
> *he has prepared his deadly weapons, making his arrows fiery shafts.*
> *Psalm 7.12-13; cf. 35.1-3; 64.7*

> *You have rebuked the nations, you have destroyed the wicked;*
> *you have blotted out their name for ever and ever.*
> *Psalm 9.5*

The examples above from Psalms 7 and 35 may be said to portray God as ready to use military methods if necessary, though there remains some hope of repentance and relief. In the cases of Psalms 9 and 64, however, there seems to be no doubt that God will use, or has already used, his military-like powers of coercion. God's use of arrows in Psalm 64.7 is almost certainly metaphorical, since it is seen as a just recompense for the enemies' use of "bitter words like arrows, shooting from ambush at the blameless" (vv3-4). Words were, in this case, the real arrows. Psalm 24.8 also depicts God as "The Lord, strong and mighty, the Lord, mighty in battle." With its overtones of procession and celebration, Psalm 24 may have been used in the context of a military victory, or a commemoration of a past victory.

Not only does God work violently in bringing judgement against Israel - the psalmists also expected God to work violently on their behalf. Psalm 10 describes the psalmist's distress and expects God to "take it into your hands ... Break the arm of the wicked" (Psalm 10.14-15; cf. 37.17), or to rain coals of fire and sulphur and a scorching wind on the wicked (Psalm 11.6; cf. 18.12ff), or to "break the teeth in their mouths" (Psalm 58.6, referring to the enemies as "young lions"), or to "shatter the heads of his enemies" (Psalm 68.21; 110.6), or to fill the lands with corpses (Psalm 110.6). Some of these violent expressions may be colourful ways of asking God to break the power or influence of the psalmist's enemies, but even so the expectation of physical violence cannot be eliminated. Further slight relief from the ferocity of these statements may be available in the thought that the psalmist is at least asking for God himself to take vengeance on the enemy and not inflicting it by the psalmist's own impetuous action, though there is a strong likelihood that the divine judgment will be applied through the action of some other human means, probably of a military nature, if not himself. When battles were frequent and hand-to-hand fighting was cruel and gory, the people's views of the ways of God would be different from those we now prefer.

The Lord is mighty in battle (Psalm 24.8) and he also "trains my hands for war" and "gave me vengeance" (Psalm 18.34, 47). In Psalm 58.10 and 68.23 the righteous are encouraged to rejoice as they "bathe their feet in the blood of the wicked", though the sense may be that of washing *off* the blood of battle. The author of Psalm 59.11 preferred a more lingering destruction of his enemies: "Do not kill them, or my people may forget; make them totter by your power, and bring them down"; the nation's appreciation of the victory will be greater if the battle is drawn out and the suffering of the defeated enemy more prolonged. Kraus argues unconvincingly that this verse is not an "inhumane impulse" but a desire that "God's intervention may be visible everywhere and of enduring effectiveness."[93] Psalms 69.19-29 and 78.1-72 are particularly noteworthy for calling upon God to act violently against his enemies, or for seeing God as active in disaster and devastation – the latter psalm is a poetic celebration of God's rejection of the northern tribes of Israel for their disobedience, and his choice of David and Judah to receive his benefits. In 78.65, God is even likened to a warrior waking from a drunken sleep, "shouting because of wine." This image serves both to recognise a previous period in which divine action seemed to be absent (while he was asleep), and to emphasise the violence of God's fury once he took the field. The nation Israel itself is not spared from these judgments: Psalm 106 describes the manner in which God's punishments on Israel were a continuing feature of their history – from Egypt, through the wilderness and into the settlement in Canaan, there seemed never to be a period when the people were not disobedient and God was not punishing them.

In the New Testament this tendency to attribute historical events to the direct action of God is maintained. The birth and ministry of Jesus are set within a framework of divine intention, announced by heavenly messengers, prefigured in the Scriptures and accompanied by miraculous signs. Jesus is able to foretell both his death and his resurrection[94] and his denial by Peter.[95] The betrayal and death of Jesus, though clearly implemented through human decisions and actions, are attributed in the apostle Peter's sermon on the day of Pentecost to "the definite plan and foreknowledge of God" (Acts 2.23; cf. "whatever your hand and your plan had predestined to take place", Acts 4.28). Prophets foretelling future events were esteemed in the early church. [96]

The above examples have demonstrated the frequent biblical understanding that God is at work in events. In some cases, for example in Jeremiah, it is recognised that God used very visible instruments – battles, invaders, disease, natural disasters – but other examples dispense with any necessity to describe an instrument. God is portrayed as the primary actor. How does this presentation of divine action accord with our understanding of historical events?

Our modern preference to interpret events in human and geo-political terms renders the biblical attribution of events to God problematical. What are the causes of wars? We might point to the opposition of competing political creeds, or the abnormal or aggressive personalities of leaders, or the need for a buffer zone between opposing ideologies, or the strategic value of a commodity such as oil. Why do epidemics occur? We know some of the conditions in which disease factors can be nourished and disseminated. Why do natural disasters happen? We can explain the forces involved in earthquakes, tsunamis, volcanic eruptions and hurricanes, even though we cannot predict exactly where and when they will occur. Our ability to analyse the reasons for events appears to render divine explanations unnecessary and unrealistic – even counter-productive, as the attribution of an event to God might disguise the physical cause and hinder work to prevent a recurrence. If all epidemics, for example, had been regarded as divine judgements, little would have been done to identify the nature of diseases and to attack their causes. If all wars had been regarded as provoked by God for righteous purposes, attempts to promote peace through dialogue and mutual understanding would not be initiated.

Our modern understanding of the causes of events can also be projected back onto the biblical story. The causes we can now identify were also present in the past, including in the events described in the biblical histories. Competing political creeds, the aberrant personalities of leaders, and the need for buffer zones or strategic products were as present then as they are now. But the biblical authors had a different perspective. We may take the view that the biblical writers were genuinely ignorant of the "true" causes of events, and could only see things in terms of actions by God. Alternatively we may consider that they were aware that events such as wars and famines had underlying causes, but they saw another set of purposes at work – divine purposes working through daily events, and it was their task to emphasise these factors.

To interpret current events as divine actions is a risky business, requiring courage and confidence. The possibility of ridicule is high. Recent examples include the competing interpretations of the devastating Indian Ocean tsunami of 2004: as an act of judgment by a Christian God upon hedonistic Europeans spending Christmas on the beaches of Thailand, or, alternatively, by a Muslim God on young Indonesians imitating Western behaviour. Both explanations are partial and inadequate. The Hebrew prophets faced similar risks in interpreting their contemporary events in terms of divine action.

A mingling of divine and human categories

We have seen above that although many of the biblical writers state firmly that God is responsible for judgments on the earth, there is also an occasional recognition of other factors at work. Some of the biblical authors display a more subtle appreciation of the experienced causes of events – instead of the simple assertion that God did X, there is a recognition that X was brought about by a decidedly earthly process, such as war, or human wisdom, or the weather, though it continues to be understood as part of God's will. These authors were aware that the reasons behind historical events were more complex than a simple assertion that God did it. Possibly they were not satisfied by the simpler affirmation, and wished to show that the will of God emerged through a great variety of factors. Those writers of the larger number of texts who were happy to state simply that God did X may have felt that that was all there was to say; or they may have been aware that more indirect processes were at work, but the nature of their storytelling or prophesy did not require them to go into historical detail. They were, after all, telling the story from the divine viewpoint. Perhaps they even felt the obligation to counteract a prevailing tendency to find causes and explanations only in historical and earth-bound factors. Some biblical writers wished to emphasise that God was the author of the process, while others wanted to show that events happened for more complex reasons.

Some examples of the more subtle mixing of both divine and human explanations may be found in a variety of the biblical sources. These are mostly books recognised as of later provenance in Hebrew history, perhaps indicating an increasing sophistication in the understanding of events. My first example, however, is from the early period, at least as far as the date of the events described is concerned (though the date of compilation of the text may be considerably later). The author of the book of Judges, having described how the rest of Israel had wreaked slaughter on the Benjaminites (Judges 20), states that "the Lord had made a breach in the tribes of Israel" (Judges 21.15). Here is *post-hoc* theologising of the most blatant kind – events that began in inter-tribal rivalry are swept up into a pious description of divine purposes.

At a later time, the author of the Books of Chronicles is aware of the need for human skills and wisdom in national leadership. The book describes those in the tribe of Issachar "who had understanding of the times, to know what Israel ought to do" (1 Chronicles 12.32); this is a more pointed view of human nature and the human contribution to events than some of the more blunt statements about divine interventions. Similarly, Chronicles describes the judges who "judge not on behalf of human beings but on the Lord's behalf; he is with you in giving judgment" (2 Chronicles 19.6-8); again, the interaction of human wisdom with some form of divine guidance is recognised. We also read about Chenaniah who "understood" music (1 Chronicles 15.22), Huram-abi who was "a skilled artisan, endowed with understanding" (2 Chronicles 2.13), and "other Levites, all skilful with instruments of music" (2 Chronicles 34.12). The downfall of Ahaziah king of Judah, captured and executed by Jehu, was nevertheless "ordained by God" (2 Chronicles 22.7), and the words even of the Egyptian king Neco were "from the mouth of God" (2 Chronicles 35.22). The Chronicler is thus attempting to show the involvement of human beings, even non-Israelites, in forwarding the purposes of God. This may be contrasted with the views expressed in Exodus 35.31-35, where practical skills are attributed firmly to the divine spirit.

The author of the book of Ezra even manages a little dry humour when he states that as the returned exiles listened in the open air to the reading of the law of the Lord, they trembled because of the seriousness of the matter and the heavy rain (Ezra 10.9). Sensibly, they appointed a sub-committee and decided to return to the issue on a later occasion. The book of Ezra is normally regarded as a harsh, unfeeling application of divine law, but in this case the author has allowed a little greyness to enter his description of events. In another example of dual explanation, we read in Nehemiah that what the king had granted was also given by the gracious hand of God (Nehemiah 2.8; 6.16).

The book of Proverbs offers a more generalised treatment of the mingling of divine and human explanations:

> *The human mind may devise many plans, but it is the purpose of the Lord that will be established.*
> *The horse is made ready for the day of battle, but the victory belongs to the Lord.*
> *Proverbs 19.21; 21.31*

These wise sayings are not suggesting that it is worthless for the human mind to devise plans, nor that there is no need to have the military machine in readiness, but that whatever the state of human preparations the outcomes are beyond complete human control, and in this must be seen the purpose of God. A more extreme view might suggest that human plans and reliance on military strength were contrary to faith in God, and that the leaders and people should trust in him *alone*. Such views were sometimes espoused by the Hebrew prophets, for example:

Oh, rebellious children, says the Lord, who carry out a plan, but not mine; who make an alliance but against my will ... the protection of Pharaoh shall become your shame ... In returning and rest you shall be saved, in quietness and in trust shall be your strength.
Isaiah 30.1-5, 15

Alas for those who go down to Egypt for help and who rely on horses, who trust in chariots because they are many and in horsemen because they are very strong, but do not look to the Holy One of Israel or consult the Lord!
Isaiah 31.1

It has been suggested that these calls to abandon alliances with Egypt were not a case of "heroic faith or heroic folly" or political "quietism" but were based on a sound assessment of particular weaknesses of Egypt during Hezekiah's reign.[97]

Also, when the city authorities in Jerusalem were planning the city's defences and water supply, they "did not look to him who did it, or have regard for him who planned it long ago", says Isaiah 22.11. These judgements in the book of Isaiah were against the failure by the authorities to see the nation in its relation to its divine ruler. Their focus was entirely upon technological or political solutions. "Israel's leaders appear to agree with the nations that the only real force lies in military strength." [98] The sayings of the wise recorded in Proverbs may have been an attempt to soften the harshness of the kind of prophetic view represented here in Isaiah (though the sayings in Proverbs were probably penned years or centuries later). It was perfectly possible for God to use earthly power or wisdom to achieve his ends, but it was necessary to accept humbly that the outcome might not be the one for which preparations had been made.

In a bold interpretation of current events, Deutero-Isaiah declares that God will make use of the Persian king Cyrus, his anointed (44.28 - 45.1), even though "you do not know me" (45.5).

Cyrus himself was not aware of the way his actions were being used in the purposes of God. The straight path in the desert, which appears at one point to be the return route to Jerusalem for Yahweh and the Israelites (40.3) was preceded by a straight path for Cyrus to conquer Israel's conquerors (45.13; 46.11). Jeremiah also claims that God works in those who do not know it: God is responsible for "planting" the wicked and enabling them to take root (Jeremiah 12.1-2). Jeremiah 13 is a chapter dominated by a sense of divine action, but in the middle of his focus on God's judgement Jeremiah recognises the human factor: "it is for the greatness of your iniquity that ... you are violated" (v22). In Zechariah's description of judgements to come, God says "I have bent Judah as my bow ... [I will] wield you like a warrior's sword" (Zechariah 9.13-15). In this case, Judah is the instrument of judgement in God's hands. Divine and human actions are intermingled and sometimes indistinguishable.

The modern interpretation of human events

How are we, with our scientific thought patterns accustomed to the inexorable flow of cause and effect, to interpret these biblical claims for the action of God in the regular events of life such as childbirth, romantic attraction, the possession of wisdom (or the lack of it), material prosperity (or the lack of it), health and healing (or their absence), and protection of every kind (or, once again, the lack of it)? And may the principle of mingled divine and human actions be applied in our own times? This question becomes more pointed when particular examples are considered. Is it possible to claim that individual events are actions by God? Imagine some randomly selected examples: the burning down of a church by vandals; a declaration of war; the unexpected death of a leader; a devastating flood. Would anyone dare to claim nowadays that events such as these were actions by God? Perhaps with long-distance hindsight it might be possible to see that the eventual outcome of a horrendous event was on balance beneficial, but to claim at the time (or even in advance) that God has performed (or is about to perform) a damaging act takes courage and some insight into divine purposes. Perhaps that is the special function of a prophet.

5

Miracles and magic

We have already seen that in the biblical faiths it is virtually axiomatic that God is a God of wonders (See "God as the cause of rare events", Chapter 4). These wonders are not only what might be called the everyday wonders of the physical world, but also special acts performed for the benefit of his people: the holding back of the sea or rivers, the collapse of city walls, sudden meteorological phenomena, devastating plagues sent on enemies and on God's own people when they disobeyed him, sudden death, speaking animals, instantaneous healings, resurrections from the dead, stopping the sun in its course, and many more. Many of these events appear to have the character of interventions, special divine interruptions of the normal physical processes.

For centuries divine interventions presented no insurmountable problem for human understanding – God was the creator and designer of the world, and he was therefore both capable of performing actions contrary to the usual course of nature and entitled to intervene when he chose. Exceptional divine actions could be regarded as the more unusual aspects of what God was always doing. But with the emergence of scientific methods and scientific ways of thought, regularity became recognised as a part of the world's operation, and divine interventions became a problem. Science assumes and depends upon the regularity of the natural processes. Observations should reveal an underlying order which can be described and in many cases mathematically formulated. Experiments should be repeatable. If enough is known about the input conditions, the output should be known. Once the mechanisms have been understood, reliable predictions should be possible. Divine interventions disturb the scientific view of order, explanation and predictability.

This desire for regularity and predictability remains an important aspect of present day science, but it is also true that science has become, mainly in the course of the twentieth century, more open to uncertainty and the possibility of unexpected occurrences. It has become apparent that many phenomena in the natural world are *stochastic* rather than *deterministic*. This means that there is a strong degree of probabilistic behaviour in the processes underlying the observed phenomena. In these cases, it can no longer be said that if you ensure A, B and C the inevitable result will be D – that would be a deterministic single-outcome process. Rather, it is now recognised that given A, B and C there is a recognised probability that D will occur. That probability may be very high (in many cases the probability will be so high that it could not be distinguished from certainty), but it remains a probability. What some people would call a miracle could therefore be regarded, with acceptable scientific rigour, as an extreme event in the probability distribution. This understanding of the probabilistic character of natural processes began early in the twentieth century through the study of the behaviour of atomic and sub-atomic matter, but has since extended to many other fields. Unusual weather conditions, spontaneous recovery from illness, exceptional rates of reproduction in particular species and spectacular waves at sea may all be understood as events towards the limits of natural probability.

However, this was not the world of the biblical writers. For them, God worked normally in expected ways, and God worked wonders in exceptional circumstances. These special events, which in the twenty-first century might be interpreted as the outer extent of probability or as happy coincidences, were seen in biblical times as divine wonders performed for a divine purpose. This acceptance of wonders and miracles led to certain features in the ways the stories were told and received.

One characteristic would be a tendency towards exaggeration. If a past event is regarded as wonderful, nothing is lost by rendering it even more wonderful. We have noted already that the Israelites' crossing of the Red Sea is attributed in one text to a strong wind blowing on shallow water, while elsewhere the sea is described poetically as a wall to the left and the right (Exodus 14.21, 22, 29; 15.8, 10). Through the manner in which a story is told or heard, natural events may become fantastic miracles. And if the storyteller is prone to exaggeration, hearers may be equally open to credulity. In our modern western societies credulity is regarded as a form of feeble-mindedness, something which critical people should have left behind. But in older times, when the boundaries between truth and fable were harder to distinguish, it was understandable that people would be ready to accept the abnormal – especially when they believed in a God who did such things. The phenomenon fed upon itself.

The miraculous events described in the Bible may be of many types. We should not necessarily apply a uniform method to the interpretation of biblical stories. Some can readily be seen as natural occurrences, beneficial or otherwise because of the particular human circumstances in which they occurred. Some may have been exaggerated in the way they were told and passed on. Others may be folk tales, which in some modern religious traditions have become actualised and treated as test cases for adherence to the strict understanding of the group. Others again, it should be recognised, have the flavour of magic. And some may indeed be special divine actions – a faith which includes the life of Jesus after his death should not rule out all possibility of miracle, though neither should it embrace too easily all claims of miracle.

The Bible is the story of the religious journey of a people, first the Jews and then the early Christians. These people believed in a God who performed wonders, and to understand the Bible requires, in the first place, that we read it from within their religious perspective. To "explain" miracles comes very close to explaining away miracles, and if we explain them away we lose a large element of the original faith. But for ourselves, with our ways of thought conditioned by centuries of science, we must necessarily take the further step of explanation, even though this will transform some of the elements of the original faith.

I do not intend to undertake here a typology of all biblical miracles, but to identify a few examples which appear to display the characteristics I have described above.

Reported miracles with the character of folk tales would be, in my view, stories of talking animals (the snake in the garden of Eden, Genesis 3.1-15, and Balaam's ass, Numbers 22.28-30), the collapse of the walls of Jericho in response to shouts and trumpet blasts (Joshua 6), the story of Jonah in the belly of a fish, and the stories of Daniel and his friends overcoming the opposition of those portrayed as God's enemies. Some of these tales may have been grafted onto traditions based on an actual historic figure or incident.

Examples of events with a "natural" origin but embellished in the telling might be the plagues of Egypt, the crossings of the Red Sea (Exodus 14) and the River Jordan (by Joshua and the people, Joshua 3-4, and later by Elijah and Elisha, 2 Kings 2.8-14), the purification of water and food supplies (by Moses, Exodus 15.23-25, and later by Elisha, 2 Kings 2.19-22; 4.39-41), and certain healings which may have included a psychological element. Accounts which include a significant role for priests and the religious symbol of the ark, such as the crossing of the River Jordan under Joshua and the capture of Jericho (Joshua 3-4 and 6) may be associated with later cultic celebrations or re-enactments.

Stories that have a distinct flavour of magic include Moses' facility to change his staff into a snake and back again, to make his hand leprous and back again, and to change water from the Nile instantaneously into blood. Although these miraculous powers are described as gifts of God (Exodus 4), their magical character is demonstrated even within the biblical account by the ability of the magicians of Egypt to perform similar acts (Exodus 7). Other events with a magical flavour are the budding of Aaron's rod to demonstrate his chosen position as priest (Numbers 17 – not merely a budding, but also blossoms and ripe almonds together overnight, v8!) and the strength of Samson existing in his hair (Judges 16.4-22). In the New Testament, the account of Jesus predicting that Peter will catch a fish with a coin in its mouth to pay the temple tax for the two of them (Matthew 17.24-27) also has the appearance of an act of magic.[99] The placing of the sick so that Peter's shadow might pass over them (Acts 5.15 – the text implies but does not actually state that such actions by the populace were effective), and Jesus' sense of a loss of power when the haemorrhaging woman merely touched his cloak (Mark 5.30; Luke 8.44; other examples occur at Mark 6.56 and Acts 19.12) would strike most modern people as examples of magical beliefs. Within the gospel accounts of the woman's healing, the religiously contagious uncleanness of the woman is contrasted with the healing that flows from Jesus' garment – but in spite of this neat contrast, the magical or superstitious nature of the transfer is still present. Jesus states that the woman's faith has made her well (Mark 5.34; Luke 8.48), the two explanations of the healing sit uncomfortably together.

Some miracles appear so contrary to known physical principles that they are particularly problematical for scientists. These include the floating axe-head (2 Kings 6.5-7), Jesus changing water into wine (John 2.1-11), and Jesus walking on the surface of the lake (Mark 6.47-52; Matthew 14.23-33; John 6.16-21). Whatever the origins of the story of his walk on the lake, its preservation in all four gospel traditions shows that it was widely accepted in the earliest church recollections of Jesus. Albright and Mann comment: "Attempts to rationalise the story, or to penetrate behind the gospel account for some natural explanation which will not offend 'modern' susceptibilities, are rather pathetic. We can never know precisely what happened on this occasion, but the spiritual import is clear."[100]

Most famous among the "impossible" miracles are the accounts of the sun standing still to give Joshua time to overcome the Amorites in battle (Joshua 10.12-14) and the retreat of the shadow on Hezekiah's sundial (2 Kings 20.8-11). Days in which much has been achieved often appear to the participants as though they have been artificially lengthened, but these two incidents are recounted in a deadpan manner as though the author accepted the literal truth of cosmic adjustments: "The sun stopped in mid-heaven, and did not hurry to set for about a whole day. There has been no day like it before or since ..." (Joshua 10.13-14). When the reasons for the sun's apparent motion through the sky were unknown, why should it be thought impossible that the sun could be halted? It is significant that the verses are described as a quotation from the unknown "Book of Jashar", and are printed in poetic verse format in modern English Bibles:

Sun, stand still at Gibeon,
and Moon, in the valley of Aijalon.'
And the sun stood still, and the moon stopped,
until the nation took vengeance on their enemies.
Joshua 10.12-13

This passage was critical in the seventeenth century trials over Galileo's claim for a moving earth – since the Scripture states that it was the sun that stood still, Galileo's opponents claimed that in normal circumstances it must be the sun which moves, and not the earth as Galileo was claiming. The mechanism of the solar system which has been established since the time of Isaac Newton tell us that the sun could remain stationary in the sky (or retreat) only if the earth stopped spinning on its axis (or began to spin backwards). The forces required to achieve such results, if they could be applied at all, would result in the disintegration of the earth.

We have noted in the discussion of dreams a tension between biblical claims of God's guidance through dreams and biblical prohibitions against the use of dreams. The same kind of ambiguity occurs in the relationships between miracles and the magical element entailed in some of these miracles. Warnings against the use of divination occur in several places[101], yet the patriarch Joseph is said to have used his drinking cup for divination (Genesis 44.5, 15). At an earlier time Leah and Rachel appear to have practised some form of sexual magic involving mandrakes (Genesis 30.14-17), and Laban too practised divination (Genesis 30.27). An internal development in the faith of Israel is the most likely explanation of these inconsistencies – the stories from the early period faithfully record certain practices which were outlawed later.

The role of what we would describe as chance or luck is also incorporated in some biblical ideas about the revelation of the will of God. Prominent in the practices of the Israelite priesthood was the use of Urim and Thummim, which were some form of decoration incorporated in the priestly garments (Exodus 28.30; Leviticus 8.8). The Urim and Thummim appear to have been a kind of yes/no means of determining the will of God – questions to God had to be asked in a form capable of a simple answer. In Numbers 27.18-21, the Urim is used in the appointment of Joshua as Moses' successor, possibly to confirm the decision about Joshua's leadership or possibly as a continuing source of instructions for the new leader. Even as late as the time of Ezra and Nehemiah, the Urim and Thummim were still in use (Ezra 2.63; Nehemiah 7.65); the context here suggests that the Urim and Thummim were used to determine the genuineness of those claiming to be priests.

In other examples, the people or leaders "inquired of the Lord".[102] Direct yes or no answers were expected, so although the Urim and Thummim are not mentioned in all these cases some form of divination or use of chance (equivalent to the heads or tails involved in the toss of a coin) was probably involved. The example in Judges 20 is particularly detailed. The Israelite armies apparently received clear instructions from the Lord on two successive days for their battle plan against their kinsmen Benjamin (vv18, 23), only to suffer heavy defeats. A third time they are given a positive instruction (v28) and this time the battle is successful. It seems that the will of the Lord, though given in clear terms, was not always straightforward in its implementation.

The use of Urim and Thummim is described at 1 Samuel 14.41-42 as a lot. Through successive eliminations, the king's son Jonathan was identified as the guilty party on that occasion. On numerous other occasions a lot was used to make decisions, though the exact mechanism of the lot is not described. In Joshua 18-19 the "casting" of lots "before the Lord" was used to determine the division of the land between the tribes, or perhaps more realistically (since the lot seems to have been limited to a yes/no answer) to confirm the decisions reached in negotiations. In Joshua 21 the lot was again used to determine the towns occupied by different families of the Levites. 1 Chronicles 24-26 describes a system of lots to determine the functions of the priests, Levites, musicians and other officials in the temple. In Judges 20.9 a lot was used to select those who would form an army. The first king Saul was apparently selected by lot (1 Samuel 10.20-21), though this seems to have been a lottery after the event, as Saul had already been anointed king (at least according to the text as presently constituted – this kind of discrepancy opens up the possibility of multiple sources for the account of Saul's appointment to the kingship). Irreconcilable traditions may be involved here – otherwise it would appear that the lottery was manipulated to give the desired result. King David seems to have used a form of lot to select those prisoners of war who would become slaves and those to be executed (2 Samuel 8.2). According to Nehemiah 10.34 and 11.1 lots were used to determine responsibility for the delivery of wood for the temple altars, and which of the returning exiles were to live in Jerusalem or in other towns. In the story of Esther, a lot was used by Haman to determine the date of his attack against the Jews (Esther 3.7; 9.24).

The book of Proverbs provides a simple Hebrew justification of the use of lots in legal disputes:

> *The one who first states a case seems right, until the other comes and cross-examines. Casting the lots puts an end to disputes and decides between powerful contenders.*
> *Proverbs 18.17-18*

In this example the casting of lots is to be seen as the necessary procedure when a case cannot be decided on the basis of the balanced arguments put forward. Why should lots be mentioned as particularly important in cases between "powerful" contenders? Perhaps the powerful on both sides would be able to employ the best legal experts deploying skilful arguments, thus making the correct decision more difficult to achieve. Murphy suggests that between powerful litigants lots would be perceived as providing a safer and more equitable outcome. [103]

The book of Proverbs goes on to claim that the Lord overrules the outcome:

> *The lot is cast into the lap, but the decision is the Lord's alone.*
> *Proverbs 16.33*

McKane suggests that the "lap" was the breast pocket in which the devices used for the lot (the Urim and Thummim?) were kept.[104] Casting the lot into an unobservable place like a pocket seems, however, a procedure unlikely to achieve satisfaction to all parties.

The use of a lot was continued among the first followers of Jesus when a lot formed part of the process to appoint a replacement disciple for the betrayer Judas. "They proposed two ... they prayed ... and they cast lots" (Acts 1.23-26). The positioning of this incident in Luke's account between the ascension of Christ and the gift of the Spirit is perhaps intended to show that the use of lots in the Christian community belonged only to this interim period.

Finally we should recognise here the occasional recognition of chance or luck in the biblical texts without the use of formal procedures such as Urim and Thummim. In Ruth 2.3 we read that "As it happened" the immigrant Ruth found herself gleaning in the field of Boaz, and we know the romantic outcome in the dynastic line which was to lead three generations later to David. The author might have made some comment about events being directed by God, but did not do so. In 1 Samuel 6 the Philistines wished to return to Israel the ark of the Lord because of the afflictions they had suffered since capturing this sacred object. Their priests and diviners insisted that the ark be returned on a new cart drawn by animals which were permitted to go where they would. If the animals made their way to Israel, "then it is he [the Lord] who has done us this great harm; but if not, then we shall know that it is not his hand that struck us; it happened to us by chance" (v9). In the event, the animals drew the cart straight to Israelite territory, in spite of their calves being shut up at home in Philistine territory. So God, not chance, was perceived to be the author of the Philistines' misfortunes, but at least the possibility of chance was recognised. The analysis of events related in this story takes place among the Philistines. The purpose of the Hebrew author may have been to point out with some irony that only Philistines could suspect that chance might be responsible for their misfortunes – 'we Hebrews knew all along that God was the cause of their troubles.'

In another incident, the manoeuvrings of David's and Saul's fighting troops in the wilderness of Maon are described (1 Samuel 23.24-28). Saul's men passed on one side of a mountain, while David's men escaped by passing on the other side. No moral is drawn in the text. The author could have added some reference to the hand of the Lord in this escape, but chose not to. The place was called Rock of Escape – the reader is left to decide whether any divine purpose lay behind this escape. Was it lucky or was God at work?

Is God Omnipotent?

The biblical descriptions of miracles, including some that on scientific principles could be described as highly improbable or even impossible, raise the question of the omnipotence of God. Omnipotent means "all powerful", and the word belongs to the same family as omniscient (all-knowing) and omnipresent (present everywhere). All these *omni-* words have been used as descriptions of God, and there may be some relation between them when used in this way, but this section is concerned with God's omnipotence only.

For many people, the question "Is God omnipotent?" will appear unnecessary and possibly irreverent. These people will respond, "Of course God is omnipotent. This is part of what it means to be God. If God were not omnipotent he would not be God, at least as far as we have known God within the Christian faith." This is an instinctive reaction which arises from centuries of Christian worship and thought about God, and of course it is a view to be respected. Nevertheless, I believe that a little more probing about the meaning of omnipotence will help us in our understanding of God and in our living with God in the contemporary world. This instinctive affirmation of the omnipotence of God is based on a significant underlying assumption about God's nature: God has to be omnipotent – it is an essential part of God's being. Our whole conception of God may be based on the idea that God is omnipotent. But are there any other reasons for thinking this, apart from what we might call religious instinct?

In the Hebrew and Christian Scriptures we find plenty of evidence about the being and nature of God which appears to affirm the omnipotence of God.

In the Hebrew Scriptures, the book of Psalms is composed of hymns, prayers, exhortations and samples of cultic liturgy from the worship of the people in the Jerusalem Temple. The present book of Psalms was probably compiled in the period between the return of the Jewish exiles from Babylon (late fifth century BCE) and a century or so before the beginning of the Christian era. The Psalms include, however, many pieces much older than the date of the compilation as a whole. In many places in the Psalms God is described as a majestic and powerful ruler over the earth. Jerusalem was the capital city and, in the pre-exilic period, the seat of the reigning king of Judah, therefore ideas about thrones and kingship would have been highly meaningful there. To describe God as a ruler and king was therefore appropriate for the people of that time and place. We must accept that using kingly language about God is to employ a metaphor – in order to convey meaning, one group of concepts relatively well understood is used to describe another group of concepts less well understood. Some of the characteristics of the known concept may be transferred to the less well known – some, *but not all*. The essence of a metaphor is that it does not map totally and exactly in every respect between the two concepts.

God was therefore described as a ruling king – this metaphor would have carried weight for the worshippers in Jerusalem, the royal city of King David and his successors. But not all the attributes of a human king were transferable to God, and vice versa. The citizens of Jerusalem, and those who wrote and conducted their Temple liturgies, probably realised this. Their kings were of varying quality: some were corrupt and unworthy leaders while all were subject to illness and death. God's kingship, even though less visible, was of course much grander than that of the one who sat on David's throne.

As we have already seen, there are many instances in the Psalms describing God as a great king enthroned in heaven and ruling over the earth, for example:

> *The Lord looks down from heaven; he sees all mankind. From where he sits enthroned he watches all the inhabitants of the earth.*
> *Psalm 33.13-14*

From this throne elevated over the whole earth, God possesses the right and the power to take any action he chooses:

> *Our God is in the heavens; he does whatever he pleases.*
> *Psalm 115.3*

According to the views of ancient Hebrews enshrined in these and other Scriptures, God is powerful in the flow of history and in the affairs of nations. The Hebrew prophets claimed to be speaking on behalf of God, and they were particularly critical of the behaviour of Israel and Judah. As a nation they had received special advantages and favours, but had not lived according to the high standards of behaviour God required. Their nationhood was based on a covenant with God which required ethical forms of living as well as ritual observances, and the nation had repeatedly and profoundly failed to satisfy the requirements of this covenant, especially in its ethical aspects. Therefore, says the prophet Amos, the nation will be punished:

> *You only have I known of all the families of the earth; therefore I will punish you for all your iniquities ... I will take you into exile ... I will deliver up the city.*
> *Amos 3.2; 5.27; 6.8*

Again as we have seen in the previous chapter, God has the power to act in this way within human affairs. According to the prophet Habakkuk and others, God is even able to stir up foreign nations and armies so that through political changes, military movements and conquest God is able to bring about the punishment of defeat and exile:

> *I am rousing the Chaldeans ... The cup in the Lord's right hand will come around to you, and shame will come upon your glory!*
> *Habakkuk 1.6; 2.16*

The "cup" in the Lord's hand is the cup of divine judgment to be drunk by the nation, a familiar image in the Scriptures. In further examples of the imagery of judgment, God's action is depicted as so real that God himself takes and uses weapons of war against those who have failed to meet his standards:

> *If one does not repent, God will whet his sword; he has bent and strung his bow;*
> *he has prepared his deadly weapons, making his arrows fiery shafts.*
> *Psalm 7.12-13*

In the above examples, drinking cups and military weapons are used as metaphors to depict the judgments which God will bring upon the nation.

It is important to note that the divine actions invoked in these statements by the Hebrew prophets occur as typical events in the historical process. Regular developments, such as the occurrence of wars, the rise and fall of nations and (in other texts) natural events such as droughts or plagues, are interpreted by the prophets as acts of God. But they are "regular" or "ordinary" events in the sense that similar things have happened many times before. There is no visible manifestation of God's presence in making these events occur. Another observer might regard them as part of the normal range of political or natural happenings. They are *seen* as actions by God by those who interpret events in a particular theocentric manner. In some cases the prophet is predicting future events, which the prophet regards as actions by God but which will subsequently unfold through regular historical processes. The reason some of the Hebrew prophets were reviled and hated by their contemporaries was not only because they were forecasting military defeat and national disaster, which undermined morale and appeared to be treachery, but also because they were claiming that these defeats were both the will of God and inflicted by God. The Hebrew people preferred to believe that God was their God in a special way, and they did not like to be told that God would turn against them.

We need to pause for a moment to assess how these examples of Hebrew testimony to the power and purposes of God bear upon the question of God's omnipotence. Undoubtedly the Psalms and the Hebrew prophets speak of God's power to influence and to pass judgment effectively upon events on the earth. God was certainly powerful, but the question of *omni*-potency raises philosophical questions that are not envisaged in the context of the Hebrew Scriptures. The fact that God's power was often understood to be displayed through regular and even unspectacular human events should in fact cause us to enquire more deeply into the nature of God's power and God's action in the world. If God's will is achieved through "ordinary" events, his power consists not in omnipotence but in the mysterious ability to ensure that those events take place. The deeds attributed by the prophets to God may tell us something about God's purposes and even his knowledge of the future, but do not necessarily bear on the question of omnipotence.

A theme frequently used in the Scriptures to demonstrate the power of God was his role in the creation of the world. A God who was able to create and sustain everything known and seen was surely powerful indeed. Numerous Psalms and other parts of the Hebrew Scriptures associate together the power of God shown in creation and what this reveals about the being of God himself, for example:

Before the mountains were brought forth,
or ever you had formed the earth and the world,
from everlasting to everlasting you are God.
Psalm 90.2

The Lord is king, he is robed in majesty; the Lord is robed, he is girded with strength
He has established the world; it shall never be moved; your throne is established from of old; you are from everlasting.
Psalm 93.1-2

For all the gods of the peoples are idols, but the Lord made the heavens.
Psalm 96.5

Know that the Lord is God.
It is he that made us, and we are his;
we are his people, and the sheep of his pasture.
Psalm 100.3

In addition to the above examples, Psalm 136 is an extended liturgy which describes God's works in creation and in the experience of Israel and draws the repeated conclusion that "his steadfast love endures for ever". The prophet Amos also saw the creative power of God extending to the stars: he describes God as "the one who made the Pleiades and Orion" (Amos 5.8).

In the writings of the prophet Jeremiah we frequently find an association between the power of God seen in creation and God's ability to work in his nation and people:

It is he who made the earth by his power,
who established the world by his wisdom,
and by his understanding stretched out the heavens ...
He makes lightnings for the rain,
and he brings out the wind from his storehouses ...
goldsmiths are all put to shame by their idols;
for their images are false, and there is no breath in them.
They are worthless, a work of delusion ...
Not like these is the Lord, the portion of Jacob,
for he is the one who formed all things,
and Israel is the tribe of his inheritance;
the Lord of hosts is his name.
Jeremiah 10.12-16[105]

It is I who by my great power and my outstretched arm have made the earth, with the people and animals that are on the earth, and I give it to whomsoever I please.
Jeremiah 27.5

Thus says the Lord, who gives the sun for light by day and the fixed order of the moon and stars for light by night,
who stirs up the sea so that its waves roar –
the Lord of hosts is his name:
If this fixed order were ever to cease from my presence, says the Lord,
then also the offspring of Israel would cease to be a nation before me for ever.
Jeremiah 31.35-36[106]

Thus says the Lord who made the earth, the Lord who formed it to establish it – the Lord is his name: Call to me and I will answer you, and will tell you great and hidden things that you have not known.
Jeremiah 33.2-3

As we have seen in the writings of the prophets, the power of God was also seen in events in history. The praise in Psalm 136 of God's actions on behalf of Israel has been noted above. Psalm 107 is an expression of praise for people who have experienced God's deliverance from assaults of various kinds, including being lost in the desert (vv 4-9), imprisonment in chains (10-16), illness (17-22) and the dangers of seafaring (23-32). According to the writer of 2 Chronicles, King Jehoshaphat knew that the outcome of an impending battle against Moabites and Ammonites was in the hands of God. In a ceremony of dedication before the battle he asked, "Are you not God in heaven? Do you not rule over all the kingdoms of the nations? In your hand are power and might, so that no one is able to withstand you" (2 Chronicles 20.6). (The "authenticity" of these words as a "genuine" statement by Jehoshaphat may be questioned, as the Books of Chronicles were compiled several centuries after the reign of that king, but for our present purposes it is sufficient that the sentiments attributed to Jehoshaphat were held by the biblical author).

This view of the power of God is foundational in the writings of the New Testament. In Matthew 10.29-30 (cf. Luke 12.6-7) Jesus says that every sparrow is known to God and even the hairs on a person's head are counted. In speaking to the Greek philosophers in Athens, the apostle Paul describes the Athenians' "unknown god" as "The God who made the world and everything in it, he who is Lord of heaven and earth" who does not need any gifts or worship from mankind as he is the giver of everything (Acts 17.24-25). In Lystra Paul again appeals to the superiority of the God who made everything and who supplies rain and food (Acts 14.15-17).

These examples confirm that God's ability to perform amazing deeds beyond human comprehension is an important part of the biblical story, whether in the Hebrew Scriptures or in the New Testament. Many of the events claimed as actions by God may have been worked through the "ordinary" processes of the natural and human world, but there can be no doubt that the texts also affirm God's ability to perform exceptional or "impossible" deeds. Some of the more dramatic include the transformation of Moses' staff into a snake and back again (Exodus 4.1-5; 7.8-12, though see comments above about the magical nature of this story), the provision of quails and manna in the desert (Exodus 16.13-36), the collapse of the walls of Jericho (Joshua 6.1-21), the miracles of Elisha including the resurrection of a possibly dead child (2 Kings 2.19-22; 4.1-6.23), and, in the New Testament the many healings performed by Jesus, the "natural" miracles of calming a storm (Matthew 8.23-27 and parallels) and walking on the lake (Matthew14.23-33 and parallels), and supremely the resurrection of Jesus himself. The apostle Paul writes of the occurrence of miracles as an expected feature in the life of the Christian church (1 Corinthians 12.10, 28-29; Galatians 3.5; cf. Hebrews 2.4).

This brief examination of biblical texts suggests that the view of God's power was that *God knows everything, and God can do anything he chooses*. Psalm 139.1-18 describes God's detailed and intimate knowledge of every person: "You know when I sit down and when I rise up; you discern my thoughts from far away." The writer could not find any place from which God was absent. God was active before the writer was born. The prophet Jeremiah states that "Nothing is too hard for you [the Lord]... I am the Lord, the God of all flesh; is anything too hard for me?" (Jeremiah 32.17, 27). According to all three synoptic gospels, Jesus also said that "for God all things are possible" (Matthew 19.26, Mark 10.27 and Luke 18.27[107] in the context of the entry of the rich into the kingdom of heaven, and Mark 14.36 during Jesus' inner struggle in the Garden of Gethsemane). In Luke's gospel the same words are used by the angel Gabriel describing Elizabeth's pregnancy in old age.

Is our examination of divine omnipotence therefore closed and complete? Has this discussion of the texts shown that for the biblical writers God was indeed the mighty source of everything, that he knew everything that was happening on the earth, and that he could take whatever actions he chose? Is divine omnipotence therefore confirmed, so that we need not take our investigation any further? I want to suggest that even with this great weight of biblical evidence there remains more to be said. Several questions still arise, and these I will now address. In particular, I will deal with the following six questions:

- Does "powerful" necessarily mean "*all*-powerful"?
- Does "God can do it" necessarily mean "God will do it"?
- Does human power and decision making limit the power of God?
- Do non-human events (or "natural" events) also limit the power of God?
- Are there other biblical themes which present a picture of God in some ways differing from the all-powerful monarch?
- How do the life and example of Jesus bear upon God's omnipotence?
- Does "powerful" necessarily mean "all-powerful"?

Accepting that God is incalculably mighty and powerful and that such characteristics have to be seen as essential features in the very nature of God, does this great power necessarily equate to omnipotence or *all*-powerfulness? The concept of *omni*-potency may be taking us beyond even the fulsome biblical understanding of God's power.

Certain logical or philosophical objections to omnipotence should be considered. Some of these objections may seem trivial, but they have to be addressed. Can even God make 2 plus 2 equal 5? Can God make a circle square, or design a plane triangle whose angles add up to more than 180 degrees? In these or similar examples we are tempted to say no, even God cannot achieve impossible things with mathematics. Some caution is necessary, however. When mathematics and science are becoming increasingly counter-intuitive in phenomena such as quantum mechanics and multi-dimensional space, we have to be reticent about affirming what is or is not possible, and a believer might say that if five thousand people can be fed by five loaves and two fishes some mysterious multiplication may be within God's power. But at the level of pure numbers and simple geometry, it does seem that God is not able to do everything.

This amounts to saying that God is limited by the realities of our present existence. 2 plus 2 must always equal 4, because those are the essential meanings of the numbers 2 and 4. To make 2 plus 2 equal to any other figure would be a contradiction, and even God, or perhaps especially God, cannot participate in a contradiction. Does this imply that there is a phenomenon called Logic or Philosophical Truth which has power over God, which is perhaps even greater than God? No, it is not necessary to draw that conclusion. Rather, we may conclude that the phenomena of the world are open to rational analysis, and one fundamental conclusion of this analysis is that both sides of a contradiction cannot be true. If we wish to express the idea theologically, we could say that God cannot contradict himself, that in God there are such things as truth and un-truth, and that his nature underlies the nature of the physical world.

In the light of this impossibility of contradictions, even the statements attributed to Jesus that "for God all things are possible" have to be understood in a more nuanced fashion. At first reading these statements appear to be a general, universally applicable affirmation of God's omnipotence. However, as in all biblical interpretation, the context of the words has to be taken into account. As we have seen, these words are used by Jesus in the context of the entry of a rich man into the kingdom of heaven. For such a person to enter God's kingdom may appear difficult, especially in view of Jesus' preceding instructions to his enquirer that he must sell all his possessions, but fortunately it is not a complete impossibility. For once, Jesus' disciples caught the hidden point by realising that one did not have to be rich to be obsessed with material things. "Who then can be saved?" they asked, and at that point Jesus reminded them that God could overcome this little difficulty.

In the Garden of Gethsemane Jesus used the same words in relation to a possible escape route from the suffering and death ahead of him. Is it not surprising that at this point Jesus invoked the power of God to do apparently "impossible" things? Could Jesus not simply have walked away from the danger, abandoned his calling, and retreated to safety in a way of life that did not threaten the civil or religious powers? Clearly he was so committed to his calling that there was no possibility of retreat. Only the infinite power of God could save him from a fate towards which he had to proceed. Such deliverance was not God's will on this occasion. God's power would be shown in Jesus' resurrection after death.

A third occurrence in the New Testament of the affirmation that "nothing will be impossible with God" occurs in the mouth of the angel Gabriel when talking to Mary of the pregnancy of the previously barren Elizabeth (Luke 1.37). Here we are in the realm of the unusual and unexpected rather than the "impossible", since Elizabeth's pregnancy was not miraculous in the sense that Mary's was as described by Luke. Perhaps the grandeur of the angel's words relates more to Mary's simple incredulity than to the possibility of a conception for a woman described as getting on in years: "Mary, don't think this is impossible, because it isn't."

I therefore conclude that there should be no difficulty in accepting that there are logical limitations on the power of God, and that the apparently universal statements that all things are possible for God become less universal when read in their textual context. These statements should be understood against a comparison with human power and what humans normally regard as achievable.

• Does "God can do it" necessarily mean "God will do it"?
The power of God to do anything (within the logical limits discussed under the previous question), even if accepted *in principle,* does not necessarily mean that God will do certain things, or that he will do them often or whenever the opportunity arises. "God can do it" is not the same as "God will do it". The history of the world tells us that God does not always do what he might have been expected to do, or do repeatedly what he reportedly did on one occasion. To take examples from the biblical story, God wonderfully delivered the people under Moses from oppression in Egypt, and this created the later belief in their permanent invincibility, but God did not do it again when the enemy was Babylon, Greece or Rome. The very existence of miracles, if accepted, demonstrates not only that God can do amazing things but also that he does not do them often – it would not be seen as a miracle if it was happening in every example of its kind.

In more recent times we have experienced many examples of apparent inaction by God when action might have been expected. Were two World Wars really necessary? If for some mysterious reason they could not be avoided, did they have to include so many ghastly incidents and such ingenious evil? Even more recently, the Boxing Day tsunami of 2004, in which tens of thousands of people with no conceivable responsibility for the event lost their lives in countries around the Indian Ocean, has become the paradigmatic example of a natural event causing human grief and devastation. Was God's power limited, or did God choose not to use it? If the latter, does he not care about human suffering? Christians will react against this last possibility. Our understanding of omnipotence has to allow for the fact that God does not always do what omnipotent love suggests he should do. This must at least place a question mark against the concept of divine omnipotence – if a loving God is powerful to avert these tragedies, but does not, it can only be concluded that God has a greater, even more loving purpose, however difficult it may be for us to conceive that possibility. And in this case, the greater purpose acts against God's omnipotence. God is limited in power by something else within God's own nature – which may be coming close to saying that God is not omnipotent at all. As we proceed with our further questions, we may begin to see what this "something else within God's own nature" may actually be.

- Does human power and decision making limit the power of God?

Human choices and decisions have some interaction with the actions of God in the world. We have seen above that many of the biblical prophets interpreted human events as actions by God. The rise and fall of kings, the outcome of battles, even the survival of nations, which on one level were determined by what we would call political or geo-political factors, were understood as divine actions fulfilling divine purposes. The prophets did not classify events as either humanly-caused or divinely-caused – their worldview was more complex. But if divine power works through human means, is it also possible that divine power may be limited by human means? Political events, both past and present, must surely persuade us that not everything that happens in the world is the will of God. Not only wars but also less violent forms of oppression, the absence of justice, devastation resulting from bad economic choices, the deprivation of basic human needs, the theft of personal freedoms, and all manner of deeds of aggression and greed can render human life less worthy and wholesome than is desirable. In such events we see the result of human decisions producing results that are contrary to the will of God. Sadly it is possible for humans to choose an alternative way. We read that in his home town Jesus could perform few acts of power because of their unbelief (Mark 6.5-6).

In theory it may have been possible for God to prevent all these awful things, and to force his perfect world upon us. But throughout human history God has not done this, and has allowed the world to develop according to human choices. Should we conclude that divine power was limited by human actions, or that God has a purpose which precludes his interference and permits human freedom, for better or for worse? In practical terms, there is no great difference between these possible conclusions. If omnipotence is only a theory, it should not be described as omnipotence – an omnipotence that is not exercised is no omnipotence at all.

- Do non-human events (or "natural" events) also limit the power of God?

Just as human decisions and choices do not necessarily embrace the will of God, but rather open the opportunity for both good and harmful events to enter the world, so also, I maintain, the non-human world includes phenomena and events that have not derived from the loving Father of Christian faith. The reasons for this lie in the freedom of action which also exists in the non-human world. In the last few decades science has come to recognise that physical processes are not totally deterministic (as was believed by some in earlier stages of the scientific endeavour) but include a significant degree of randomness – this randomness can be understood as a form of freedom in the non-human world. The mutation of genes, the changes in the physical environment, and the emergence of new species all depend on random variations. Some freedom of choice also exists in the behaviour of the non-human animal world, admittedly to only a small extent in many species, but increasingly at the levels of higher species.

Within limits, animals can choose where to move, when and

what to eat, and when and with which other member of their species to mate. This freedom introduces a wonderful richness and variety into the animal world, but it also leads to pain and suffering as carnivores hunt, kill and consume their prey. The methods of reproduction of some parasitic insects inevitably cause pain and death to their hosts. Animals may lack the cognitive ability to anticipate pain and death, which therefore do not hold the same fears as experienced by humans, but there can be no doubt that animals experience pain and attempt to avoid it. The non-human world cannot be measured according to human ethical standards, but fear, pain, aggression and even cruelty are not restricted to humans. The idea that the non human world is full of peace and pristine perfection, and that only the human world contains nastiness and conflict, is romantic nonsense.

The phenomenon of randomness exists throughout the physical world, whether in the behaviour of sub-atomic particles at the quantum level, or in the turbulence of air masses and oceans which determines our weather. And if randomness is genuinely random, even though working within a stable system of physical laws, it is difficult to conclude that the outcome accords with some intentional plan, or with the will of God. It appears that the non-human or "natural" world operates in a manner which restricts any concept of divine omnipotence.

- Are there other biblical themes which present a picture of God in some ways differing from the all-powerful monarch?

I have argued in the opening part of this section that the Hebrew and Christian Scriptures provide an understanding of God as enormously powerful and knowing, even if this does not extend to omnipotence in the full philosophical sense. But is this the only view of God present in the Scriptures? Are there additional insights into the nature of God which have to be set alongside the image of a powerful monarch?

In the biblical story and since, the human appeal for help and deliverance through God's almighty power has not always been answered. The assurance that things will be all right "in the end" may be a necessary ingredient of faith, but in the meantime suffering and uncertainty continue while God appears to be absent. Before the deliverance of the Israelites from Egypt under Moses generations of Israelites were exploited and suffered, but God did not rescue them at that time. For the nations of Judah and Israel there was constant conflict with warring neighbours great and small. Facing these disasters, many of the Hebrew prophets emphasised the importance of righteousness in following the covenant between God and Israel: the nation could not expect God to work for them if they did not observe his ethical requirements. According to the prophets, this failure was the reason for God's distance and judgement upon them. However, the Book of Job challenges this view, at least in its application to individual people, by exploring the suffering and loss which afflict even the righteous and God-fearing.[108] One prophet concluded that "Truly, you are a God who hides himself" (Isaiah 45.15).

Several Psalms witness to the individual's sense of loneliness and despair when God seems far away, for example:

Why, O Lord, do you stand far off?
Why do you hide yourself in times of trouble?
Psalm 10.1

How long, O Lord? Will you forget me for ever?
How long will you hide your face from me?
How long must I bear pain in my soul, and have sorrow in my heart all day long?
How long shall my enemy be exalted over me?
Psalm 13.1-2

My God, my God, why have you forsaken me?

Why are you so far from helping me, from the words of my groaning?
O my God, I cry by day, but you do not answer; and by night, but find no rest.
Psalm 22.1-2

I am weary with my crying; my throat is parched.
My eyes grow dim with waiting for my God.
Psalm 69.3

How long, O God, is the foe to scoff?
Is the enemy to revile your name for ever?
Why do you hold back your hand; why do you keep your hand in your bosom?
Psalm 74.10-11

How long, O Lord? Will you hide yourself for ever?
Psalm 89.46

The prophet Hosea found complex parallels between his own agonising attachment to an unfaithful wife (or, on one interpretation of his work, wives) and God's relationship with Israel. Both relationships contained love, generous nurturing, passion, betrayal and grief, and Hosea portrays God as torn and confused in his love for an unresponsive nation:

How can I give you up, Ephraim? How can I hand you over, O Israel? ...
My heart recoils within me; my compassion grows warm and tender.
I will not execute my fierce anger; I will not again destroy Ephraim.
Hosea 11.8-9

In portraying God as a love-smitten loser like himself, unable to inflict the punishment the nation deserves, Hosea's understanding of God is a million miles from divine omnipotence. God and Hosea are in the same position: overflowing with a love that is returned with faithlessness, and able only to renew love again and again. This is a view of God very different from the all-powerful ruler and king.

- How do the life and example of Jesus bear upon God's omnipotence?

For Christians, Jesus Christ is the visible image of God. "He is the reflection of God's glory and the exact imprint of God's very being" (Hebrews 1.3). "He is the image of the invisible God" (Colossians 1.15). Difficult to grasp as these statements are, they nevertheless encapsulate the Christian belief that in some way the infinite greatness and majesty of God was expressed in that human life. Among the many remarkable results of this belief is that it brings human life and divine life much closer than in many other systems of belief. We therefore look at the life of Jesus Christ to find some clues about divine omnipotence.

In the accounts of Christ's life, we are immediately struck by his poverty, his humility and his defencelessness in face of death. From the stories of an animal manger being used as his first cradle, through his youthful life as the working son in a peasant village, to the years of his itinerant ministry when he had "nowhere to lay his head" (Matthew 8.20; Luke 9.58), we encounter one who owned little or nothing, who shared a common purse with his band of followers, and who enjoyed a good meal but otherwise seemed to care nothing for the symbols of comfort and success. Towards those in whom he detected hypocrisy he could be biting in his speech, but towards the honest and repentant he was tender and forgiving, and he even tolerated an enemy within his close circle. He accepted into himself the violence of a brutal flogging followed by crucifixion. The apostle Paul, probably using an early Christian liturgy, wrote:

> *Though he was in the form of God ...*
> *[he] emptied himself, taking the form of a slave, being born in human likeness.*
> *And being found in human form, he humbled himself and became obedient to the point of death – even death on a cross.*
> *Philippians 2.6-7*

If this is the "imprint of God's very being", what does it tell us about divine omnipotence? At the very least, omnipotence is deeply suppressed; possibly it is not present at all.

The occasional flash of anger, such as when he cleared tradespeople from the temple courtyard[109], was perhaps one basis for the early Christian church's view of Christ as the eventual judge as well as saviour of the world. The concept of a final day when all will be revealed and the guilty will be judged is strongly present in the New Testament faith. In John's gospel we are told that God did not send the Son into the world to condemn the world but to be its saviour (3.17; 12.47). Nevertheless, in the same gospel the Father "has given all judgement to the Son" (5.22, 27), and the words spoken by the Son will be the criteria of judgement (12.48). In Matthew's gospel the parable of the sheep and the goats depicts the Son of Man, a favourite title for Jesus, as the judge on the throne (Matthew 25.31-46). In the preaching of Peter and Paul in the book of Acts, Jesus is described as "the one ordained by God as judge of the living and the dead" (Acts 10.42; cf. 17.31). In Romans 2.16 Paul wrote that "God, through Jesus Christ, will judge the secret thoughts of all", and in 2 Corinthians 5.10 the final judgement is said to be before "the judgement seat of Christ". A violent image of Christ the Word of God is present in the Book of Revelation, where Christ strikes down the nations by the sword which comes from his mouth (19.15, drawing on Isaiah 11.4 and 49.2) – a warlike picture, but nevertheless making the same point as John the evangelist, that it is the words of Christ that provide the measure against which judgement will be given.

The New Testament picture of Christ is, therefore, not all tenderness and love (though a degree of distinction between the synoptic gospels and the later writings of the church has to be noted). It does appear that the role of Christ as judge is reserved for the final summing up of all things, and this gives us a further insight into divine omnipotence. If God were to exercise his omnipotence by putting right all wrongs, and introducing the kingdom as he intends it, that would indeed be the beginning of a new world. The rules would be changed and perfect completion would have arrived. In the meantime, things are not like that. God does not intervene too often, but he allows the world to make its choices and to contribute to the development of God's intentions. God limits his own omnipotence because he desires the world to grow towards his perfection.

God's greater purpose is that the world should move freely towards his own perfection – this is the "something else within God's own nature" that we have noted above, the element in God that restricts the operation of his omnipotence.

Science, miracles and magic

At one level, scientific endeavour is based upon the search for order, regularity and explanation through established physical principles. When considered at this level, a scientific approach will have difficulty with both miracles and magic, as they appear to defy the principles of cause and effect that are so important to scientific explanation. However, at another level science has learned during the twentieth century to accept the unusual and unexplained, and even the contradictory, as part of the evidence from the natural world that has to be included within a genuinely comprehensive system. Science, having once been regarded as a supremely logical set of disciplines, has through the emergence since the 1920s of quantum mechanics, Heisenberg uncertainty and chaos theory come to be regarded as in some ways illogical and counter-intuitive. There is room near the limits of probability for rare and unexpected events.

This does not mean that all the biblical accounts of the miraculous will be accepted at their face value. Critical questions will still be asked, and factors such as narrative exaggeration and theological bias will be taken into account.

Although miracles and magic have been treated together in this chapter, it is also necessary at this stage to distinguish between them. While the miraculous focuses on the result or outcome, usually beneficial, and leaves the process unexplained, the magical places its emphasis on the method by which something abnormal is achieved. While miracles create a sense of awe, the magical can produce the desire to possess and copy a technique. At its best magic may be no more than innocent deception, while at its worst it is chicanery or even evil.

6

Science and beyond science

Is there some science in the Bible after all?

In previous chapters we have examined some of the ways in which the outlook of the biblical writers on the natural world or on human experiences are different from, even in opposition to, the outlook of science. I have argued that in many respects it is necessary for the present-day reader of the Bible to recognise the biblical perspective as limited and in some senses incorrect in the light of more recent knowledge, though it is also necessary to understand the biblical perspective in order to appreciate well the meaning of the text. I now turn to examine whether, in some respects, the Bible does in fact contain material which may be regarded as evidence of an outlook which could be cautiously described as "scientific".

I insist at the outset that that I am emphatically *not* attempting to demonstrate that the biblical writers had some hidden knowledge of later scientific findings – whether this was secret knowledge which the biblical writers themselves possessed but hid from their audiences, even though, according to this view, occasional glimpses of their secret knowledge can be detected; or whether this was divine knowledge hidden from the writers themselves but esoterically hidden within the text. Thus I am *not* arguing, for example, that the statement in Isaiah that God "sits above the circle of the earth" (Isaiah 40.22) is evidence that the prophet knew that the earth was a sphere, nor that God was giving here a hint that God knew the earth was a sphere. This text can be readily explained on the basis of the circular panorama from any elevated vantage point. Nor am I remotely suggesting that the days of creation described in Genesis 1 reproduce so exactly the chronology of the earth's geological development produced in the nineteenth and twentieth centuries (in fact, they do not) that the author of Genesis 1 must have had a God-given knowledge of the true history of the earth. These verses, and others which have sometimes been claimed as indications of secret advance knowledge of findings not discovered by science until centuries later, are better explained in terms of their own cultures and beliefs at the time they were written (or, perhaps, of the even earlier times from which the authors were drawing parts of their material).

Rather, what I wish to explore here is whether the biblical writings provide evidence that in the societies from which the Bible emerged there existed mental attitudes and a sense of enquiry which produced knowledge which later would be described as of a "scientific" type. The scientific method is particularly concerned with the investigation of causes – can we find in the Bible writings any sustained knowledge of material causes of material events, or were the Bible authors always inclined to attribute everything that happened to the direct action of God?

The search for material causes takes place through several different types of activity. Scientific knowledge advances in an iterative, cumulative manner. The essential steps may be described as Observation, Experiment, Theoretical Explanation and Prediction. Although these can be distinguished as distinct steps in the investigative process, they are frequently interlaced – progress with one type of activity may lead to a revision and repetition of earlier stages of the process. The distinction between Observation and Experiment is that the former relies solely on detecting and remembering what occurs in the natural world, while the latter aims to test the effects of systematically imposed variations. Experiments do not have to be conducted in laboratories; they may take place in the field, literally, by testing different methods of farming or horticulture.

An initial assessment would suggest that while the Bible contains very many examples of Observation of the natural world, there is little evidence of Experiment and even less of Theoretical Explanation or Prediction. While broadly true, this assessment itself requires some careful examination.

The *Observation* of natural phenomena is, not surprisingly for a nation mainly rooted in agriculture and a rural economy, very common. Some of the images are picturesque and colourful for modern readers with little contact with ancient Near Eastern culture or the countryside in general.[110] The wisdom books, particularly Psalms and Proverbs, are full of natural images based on careful observation, of which a few examples follow:

They [those who follow the law of the Lord] are like trees planted by streams of water, which yield their fruit in its season, and their leaves do not wither.
Psalm 1.3

In the environment of the Middle East the need of trees for a plentiful supply of water is more apparent than in more humid climates. In well-watered regions trees may grow almost anywhere by tapping into damp ground and even by drawing moisture from the damp atmosphere. In arid regions, however, trees can prosper only in close proximity to permanent water-courses, giving rise to the appearance of green ribbons through dry terrain.

As a deer longs for flowing streams.
Psalm 42.1

Like the deaf adder that stops its ear, so that it does not hear the voice of the charmers or of the cunning enchanter.
Psalm 58.4-5

Let them [the wicked] vanish like water that runs away; like grass let them be trodden down and wither. Let them be like the snail that

dissolves into slime; like the untimely birth that never sees the sun.
Psalm 58.7-8

Here in Psalm 58 reference is made to observations of four phenomena that rapidly disappear – water draining quickly into dry ground, grass trodden down along a pathway, the snail which, it was believed, dissolved into its own trail, and the miscarried foetus that does not enter the world of the living.

As smoke is driven away ... as wax melts before the fire.
Psalm 68.2

The insubstantial nature of smoke or melted wax is used here to emphasise the weakness of God's enemies.

Go to the ant, you lazybones; consider its ways, and be wise. Without having any chief or officer or ruler, it prepares its food in summer, and gathers its sustenance in harvest.
Proverbs 6.6-8

Like vinegar to the teeth, and smoke to the eyes, so are the lazy to their employers. Proverbs 10.26
The beginning of strife is like letting out water [urine?]*; so stop before the quarrel breaks out.*
Proverbs 17.14

Like vinegar on a wound is one who sings songs to a heavy heart.
Like a moth in clothing or a worm in wood, sorrow gnaws at the human heart.
Proverbs 25.20

Like a sparrow in its flitting, like a swallow in its flying, an undeserved curse goes nowhere.
Proverbs 26.2

Use of the verb "flitting" gives a good impression of the apparently directionless motion of some small birds, in agreement with the proverb's main point that an undeserved curse will be ineffective. The application of the same idea to the more elegant and purposeful flight of the swallow seems rather more ill-placed – was it not evident that the darting movements of the swallow were in pursuit of insects for food?

Like the glaze covering an earthen vessel are smooth lips with an evil heart.
Proverbs 26.23

A stone is heavy, and sand is weighty; but a fool's provocation is heavier than both.
Proverbs 27.3

Iron sharpens iron, and one person sharpens the wits of another ... Just as water reflects the face, so one human heart reflects another.
Proverbs 27.17, 19

These last two sayings, drawing on observations of the very different natural phenomena of the ability to sharpen iron tools on another piece of iron and the reflective qualities of a water surface, draw the similar conclusion that through relationships with other people we come to a better understanding of ourselves.

Can papyrus grow where there is no marsh? Can reeds flourish where there is no water?
Job 8.11

Regardless of the point the writer is making, which is not immediately apparent, this verse provides a clear example of the observation and identification of cause and effect.

The Preacher of Ecclesiastes sees in the regularities of the natural world the basis for his weariness with life. Far from being inspired by the features of the world he observes, he seems to find their constant repetition monotonous. Generations come and go, the sun rushes across the sky to return to its resting place, the wind goes round and around, and the water flows endlessly from the streams to the sea:

> *A generation goes, and a generation comes, but the earth remains for ever.*
> *The sun rises and the sun goes down, and hurries to the place where it rises.*
> *The wind blows to the south, and goes round to the north;*
> *round and round goes the wind, and on its circuits the wind returns.*
> *All streams run to the sea, but the sea is not full;*
> *to the place where the streams flow, there they continue to flow.*
> *All things are wearisome.*
> *Ecclesiastes 1.4-8*

Another set of quasi-scientific observations found in the wisdom literature of the Hebrew Scriptures is the classification of related phenomena. This may be regarded as a part of the scientific task of observation, and attempted classifications are evident in the poetic descriptions in Proverbs 30 of "three things/four things" which behave in similar ways:

> *Three things are never satisfied; four never say, 'Enough': Sheol, the barren womb, the earth ever thirsty for water, and the fire that never says, 'Enough'.*
> *Proverbs 30.15-16*

> *Three things are too wonderful for me; four I do not understand: they way of an eagle in the sky, the way of a snake on the rock, the way of a ship on the high seas, and the way of a man with a girl.*
> *Proverbs 30.18-19*

Under three things the earth trembles; under four it cannot bear up: a slave when he becomes king, and a fool when glutted with food; an unloved woman when she gets a husband, and a maid when she succeeds her mistress.
Proverbs 30.21-23

Four things on earth are small, yet they are exceedingly wise: the ants ... the badgers ... the locusts ... the lizard.
Proverbs 30.24-28

Three things are stately in their stride; four are stately in their gait: the lion ... the strutting rooster, the he-goat, and a king striding before his people.
Proverbs 30.29-31

The three things/four things literary style has been identified in other near Eastern literature. It could be taken to mean that the number is indefinite – three or four examples are given, but there may be more. The verses in Proverbs 30 may also be the answers to didactic riddles: tell me three or four things that are never satisfied/beyond understanding/under which the earth trembles, etc. It has also been suggested that the style is intended to draw attention to the last example in the set, though this explanation is not persuasive in all cases.[111] We may enjoy the wit in these classifications, especially the comparisons of human and animal behaviour and the suggestion that animals often display greater skill and wisdom, but as evidence of proto-scientific thought they do not progress beyond the category of observations of natural phenomena.

Chapters 38 to 41 of the Book of Job contain some of the most impressive observations of nature in the Bible, indeed in ancient literature as a whole, and I have already referred to some of its verses in earlier sections of this book. The chapters ostensibly consist of God's challenge to Job after the poor suffering man has pleaded his case that he does not deserve the terrible crises he has experienced. In his speech God brings forward some of the awesome aspects of the natural world to silence Job's arguments and to convince him that he, Job, has only a tiny understanding of what is happening in the world. Job's questions about his sufferings are not answered in God's speech. The implication, not stated expressly, is that his sufferings are understood by God and are a necessary part of some larger picture. Job's response is "I lay my hand on my mouth" and "I despise myself, and repent in dust and ashes" (40.4; 42.6).

Although presented as a sermon by God, these chapters were of course written by some (unknown) human being, and tell us about the author's observations of nature, which he (it is a reasonable presumption that in Hebrew society the author would be male) regarded as sufficiently wonderful and incomprehensible to be used as God's put-down argument to bring the contentious Job to a humbler frame of mind. Among the themes used in the speech are the mystery of creation itself, when no human being was present to observe what was happening, nor to help God in the task; the formations of the stars; the faithful repetition of day and night; the unknown size of the earth; the limits placed on the sea; the processes of the weather producing snow, hail, wind, rain and ice; the activities of wild animals including lions, mountain goats, wild asses and oxen, ostriches, horses, hawks and eagles; and the mysterious beasts Behemoth and Leviathan, which sound like the hippopotamus and the crocodile with a few mythological features added. The author had seen and learned much about the natural world, but remained in awe of these grand and unexplainable phenomena. In our time, we know much more about these aspects of nature, and the argument in these chapters is therefore not as convincing as it was in the centuries BCE, but the awe in face of natural beauty and sustained activity should remain.

The Hebrew prophets also observed in the natural world

analogies for, or contrasts with, their primary concerns about human behaviour in relation to God's requirements. The book of Hosea is particularly rich in figures drawn from both the natural world and the human world. In Hosea's works God is likened to a lion, a leopard, a she-bear, rain and dew, a tree, a physician, a shepherd and a hunter. Israel is likened to a herd, a heifer, grapes and a vine, figs, a fickle bird, morning mist and dew, an invalid, a mother in labour, an unborn child, an aging man, a false trader, an unturned loaf of bread, and a sown field. [112] The weak king is likened to a twig carried away by a stream. The prophet Isaiah observes that the behaviour of humans does not display the reliability and constancy of domestic animals: "The ox knows its owner, and the donkey its master's crib; but Israel does not know, my people do not understand" (Isaiah 1.3). The disciplines of ploughing and sowing, and the differing tools used for threshing different types of grain, are described at Isaiah 28.24-29, where accumulated agricultural knowledge is regarded as of divine origin: "they are well instructed; their God teaches them ... this [knowledge] comes from the Lord" (vv26, 29). Deutero-Isaiah knows that the water from rain and snow are essential for the growth of crops and uses this process as an image of the outgoing, effective word from God's mouth (Isaiah 55.10-11), and Jeremiah notes that while the migrations of the birds follow specific sequences (Jeremiah 8.7) the people by contrast do not observe God's ordinances. Another observation by Jeremiah is that the partridge hatches eggs that it did not lay (17.11) – the hatchlings will revert to their true type, and in the same way wealth acquired unjustly will, he says, quickly disappear.

In the New Testament, the parables and teachings of Jesus are justifiably regarded for their observation of the natural world and the life of a rural or small town economy. Flowers, trees and birds, sowing and harvesting, sheep and goats and shepherding, managing a vineyard, trading, fishing, patching worn clothes, weddings and funerals, robbers and cheats, kings and feasts, the taste of salt – all appear in his teaching. Acute observation of the world around was characteristic of Jesus' didactic style.

The book of James also contains numerous similes based partly on observation of the natural, animal or human worlds, and partly on themes in the Hebrew Scriptures and in the teaching of Jesus. Doubting believers are likened to waves of the sea, driven and tossed by external influences (James 1.6). The rich and their riches will disappear like a flower whose short-lived beauty perishes under the sun (1.10-11; in this case the background may lie in similar words at Isaiah 40.6-7 or Matthew 6.28-30 rather than any observation by James). The examples of bits in the mouths of horses or small rudders steering huge ships are used to demonstrate the importance of the human tongue, such a small part of the body yet capable of spreading damage like that of a forest fire (3.2-6). Animal species can be tamed (though perhaps not "every" species, including reptiles and sea creatures, as claimed by James at 3.7; his statement may be based on a hopeful anticipation of the command in Genesis 1.26-28 that mankind should have dominion over the fish of the sea and all the wild animals and every living thing). A spring cannot produce both fresh and brackish water, nor a fruit tree figs and olives and grapes (3.11-12, similar to Matthew 7.16 and Luke 6.43-44). "You are a mist" writes James, in language reminiscent of Hosea, the great Hebrew prophet of natural observation (James 4.14, Hosea 13.3). James appeals for patience in the Christian's life, on the basis of the patience of the farmer as he waits for his crop (5.7).

The second letter of Peter contains a single example of natural observation, in its reference to the unsatisfying nature of some false teachers, "waterless springs and mists driven by a storm" (2 Peter 2.17) – the image of water in satisfying human needs is here applied to the need for reliable teaching about righteous life and behaviour. In similar fashion the letter of Jude likens false teachers to irrational animals responding only to instincts, and to waterless clouds driven by the winds, trees without fruit, wild waves of the sea, and wandering stars (Jude 10-13). The reference to "wild waves of the sea, casting up the foam of their own shame" may draw on Isaiah 57.20, "the wicked are like the tossing sea that cannot keep still; its waters toss up mire and mud", while the "wandering stars" may refer to the fallen angels in 1 Enoch.[113]

The same interest in natural phenomena cannot be attributed to most of the rest of the New Testament, which only rarely makes any reference to the natural world – and even then the observation may be mediated through the Scriptures rather than based on personal experience. As we have seen in the section *God, animals and birds* in Chapter 3, the apostle Paul in 1 Corinthians 9.9 used Deuteronomy 25.4, "You shall not muzzle an ox while it is treading out the grain", merely to support his argument that church leaders should be entitled to material rewards for their work. The text was written (or, indeed, spoken by God) "entirely for our sake" This disappointing reference to the animal world seems to arise solely from Paul's knowledge of the Scriptures not from any observation or firsthand knowledge of farming.

The practice of Observation of the natural world is thus richly present, at least in some of the authors of the Hebrew and New Testament Scriptures and in Jesus himself. What can be said of the second aspect of scientific endeavour, namely *Experiment*?

One area in which the ancient Hebrews do appear to have accumulated some rudimentary scientific knowledge is in the area of human diseases and health protection. Leviticus 13-15 provides a kind of health-care handbook for the diagnosis and treatment of skin diseases and moulds, and while these chapters are presented as commands of God to Moses and Aaron (Leviticus 13.1) we may assume that the knowledge arose from trial-and-error experience in the treatment of diseases in a close-knit community. The chapters cover diseases of the skin, the onset of baldness, fungi and moulds in cloth and animal skins, the spread of moulds in houses, and bodily discharges. In cases of skin disease, the condition of bodily hairs within the area of the disease, and the depth to which the disease had apparently penetrated, were particularly important diagnostic features, while periodic isolation and washing appear to have been the only measures of control. The washing of hands is specifically mentioned, and earthen vessels which had come into contact with discharges were to be destroyed (Lev 15.11-12). The requirements for washing appear in these verses to have served a genuinely hygienic rather than a ritual purpose, though the ceremonial sense came to dominate later practice. Much benefit to human health might have been achieved if the hygienic nature of these precautions had been more thoroughly observed in succeeding centuries. In the time of the New Testament rigorous washing of hands, food and utensils was practised by "the Pharisees and all the Jews" (Mark 7.4), but in his opposition to superficial rituals Jesus seems to have criticised this practice: "to eat with unwashed hands does not defile" (Matthew 15.20).

Little information is provided about medical treatments used in biblical times, though doubtless these were common. Isaiah prescribed figs for Hezekiah's boil (2 Kings 20.7), though the Chronicler was less well disposed to medical interventions. We read that "In the thirty-ninth year of his reign Asa was diseased in his feet, and his disease became severe; yet even in his disease he did not seek the Lord, but sought help from physicians. Then Asa slept with his ancestors, dying in the forty-first year of his reign" (2 Chronicles 16.12-13). One can almost hear the Chronicler gloating over Asa's death after two years of suffering while relying on doctors rather than the Lord.

Further examples of hygienic practices are found in the requirement to consume the permitted parts of sacrifices before they had time to become rotten (Leviticus 7.15-18; 19.6-7; 22.29-30), the isolation of persons and vessels which had been in the presence of a dead body (Numbers 19.14-16), and the location of toilet facilities "outside the camp" (Deuteronomy 23.12-14). Ritual cleanliness was a particular requirement for a military camp engaged in the holy war, but the context in Deuteronomy 23 appears not to be restricted to military operations. Even the Passover meal had to be fully consumed, with none left until the next day (Numbers 9.12). While the stated motive for the early consumption of sacrificial meat is that it should be "acceptable" to the Lord (Leviticus 7.18; 19.5, 7), we may assume that the avoidance of decay and disease was the underlying consideration. The preferences of God in relation to meat were believed to coincide with human needs for hygiene. One noteworthy feature of the provision of the manna in the wilderness is described in Exodus 16.19-25: the manna collected on the sixth day was said to have remained fit for consumption on the Sabbath, whereas the manna collected on the other days had to be eaten on the same day.

Another area in which some kind of rudimentary knowledge from experience appears to have been obtained was in relation to the classifications of animals, fish and birds which could or could not be eaten (Leviticus 11, Deuteronomy 14.4-21). Of land animals, only those which both chewed the cud and had divided hoofs were to be eaten. Among the fish, only those with fins and scales could be eaten. No similar defining physical criteria appear to have been available for the birds: an itemised list of prohibited birds had to suffice, though many of those prohibited are birds of prey or feed on carrion, which might render them carriers of disease (Lev 11.13-19). Winged insects that walk on all fours with jointed legs were also permitted for food. Some of the restrictions on eating animals may appear arbitrary, but they possibly represented good hygienic practice. On the other hand, the prohibited animals may have been associated with the religious practices of some of Israel's neighbours. Since the killing of animals for food included the taking of life and was therefore accompanied by religious ritual, there could have been a fear that heterodox practices would enter Israel's religion if certain animals, possibly revered by other cultures, were killed for consumption.

Strict requirements covered not only the consumption of the animals themselves, but also the need to purify any clothing, utensils or even seeds which came into contact with a prohibited animal. In the case of seeds, dry seed which experienced contact with a prohibited animal remained clean, but damp seeds affected in this way became unclean (Leviticus 11.37-38) – it seems to have been appreciated that damp seeds might have already started the process of germination, and were thus regarded as open to infection. Water cisterns, however, remained clean even when the carcass of a prohibited animal had fallen in; somewhat illogically, anything that touched the carcass in the cistern became unclean (Leviticus 11.36), suggesting that any knowledge of the transfer of infection was limited. Water cisterns were, perhaps, too valuable to be subject to cleanliness laws which required the destruction of the affected item. It is suggested by Hartley that water being continuously renewed was regarded as a cleaning agent.[114] Perhaps the invisibility of infection spreading through water was also significant.

The obscure law against mixed fibres (Deuteronomy 22.11) would be enormously prohibitive in relation to many twenty-first century fabrics. The immediate context in Deuteronomy also forbids other mixed activities: a second crop should not be planted in a vineyard, and an ox and a donkey should not be yoked together. Planting two crops together might have been prohibited because of a fear of the spread of plant disease, while the yoking together of an ox and donkey could have serious disadvantages for the weaker animal. Reasons for outlawing mixed fibres are more difficult to identify – perhaps it was based on a general desire for simplicity and wholeness. Alternatively, mixed fibres may have been connected to the undesirable production of sweat, which was regarded as offensive to God (see, from a later period, Ezekiel 44.18). Priests were required to wear linen garments throughout (Exodus 28; 39.27-29; Leviticus 6.10; 16.4). Other interpretations of the prohibition of mixed fibres include possible associations with garments used by other peoples or worn by prostitutes. Christensen notes that even Jewish rabbis were unable to find an explanation. [115]

A further area in which experienced-based knowledge appears to have been used relates to incest and the permitted degrees of marriage (Leviticus 18.6-23; 20.10-21). Sexual relations within close degrees of relationship would not only undermine the stability of families but would also lead to increased risks of deformity in offspring.

The above examples have shown that learning from experience was important in several aspects of the life of the Hebrews. I have, however, found little evidence within the Bible for what might be described as "experimental" agricultural practices, but it is difficult to imagine that they did not occur. Farmers will have discovered which sites and soils were suited to particular crops, or the optimal times for sowing or harvesting. Any agricultural society will have developed rules and traditions for these things, based on experience. The writer of the proverb knew the rather obvious truth that "for lack of wood the fire goes out" (Proverbs 26.20) and we may take it as certain that the Hebrews would have learned and preserved knowledge about many practicalities of agricultural and small town life. As already noted, Isaiah 28.27-28 describes the different tools and techniques required for threshing different crops. The saying that "The righteous know the needs of their animals" (Proverbs 12.10) reveals an appreciation of the careful and humanitarian observation of animal behaviour; this is one of the few biblical references to the desirability of the humane treatment of animals. And Jesus told a story of the dedicated gardener pleading with the owner to give the unfruitful vine one more chance: "Sir, let it alone for one more year, until I dig round it and put manure on it" (Luke 13.8). Such common agricultural practices must have been discovered and practised, though the process of experiment and discovery is not described in the biblical texts, which have entirely different purposes.

Similarly, knowledge based on experimentation must have been available in other areas in addition to agriculture: in cooking and nutrition, in building and construction, and in the manufacture of implements and furnishings. But again, little reference to this kind of everyday experience is recorded in the Bible.

There is even less evidence of the third and fourth elements of scientific activity, *Theoretical Explanation* and *Prediction*. The biblical writers and their fellow citizens knew the normal courses of events, but no attempt appears to have been made to explain the processes involved. Some simple interpretations of weather patterns appear to represent the limits of theoretical speculation: "The north wind produces rain" (Proverbs 25.23) [116], or "When clouds are full, they empty rain on the earth" (Ecclesiastes 11.3). But even these forecasting methods are an unreliable guide to farmers' decisions: "Whoever observes the wind will not sow; and whoever regards the clouds will not reap" (Ecclesiastes 11.4). Sometimes one has to go ahead with what seems best, in spite of the surrounding indications. Murphy writes that this saying is aimed at "those who are paralyzed by their concern for the right time."[117] In general, "the work of God, who makes everything" cannot be found out (Ecclesiastes 3.11, 8.17; 11.5). The Preacher is even unconcerned about the direction in which a tree may fall, even though this is an event over which some human influence could be exerted: "whether a tree falls to the south or to the north, in the place where the tree falls, there it will lie" (Ecclesiastes 11.3). Jesus did not disagree with attempts at short-term weather forecasting, based on the appearance of the sky or the direction of the wind (Luke 12.54-56; the similar though verbally different text at Matthew 16.2-4 is absent in several ancient versions), though in the same passage he pointed out that his hearers were less adept at interpreting current events.

Exceptional events were even less open to theoretical explanation. Special events accompanying the lives of Moses and Aaron, and Elijah and Elisha, were preserved in the tradition and were regarded as special actions of God for which further explanation was not possible.

Scientific activity is composed of Observation, Experiment, Theoretical Explanation and Prediction. I have shown that although the biblical cultures were not scientific in the modern sense, they displayed some characteristics based on the observation of the natural world and learning from experience. Parts of the Hebrew Scriptures and the New Testament include acute observations of nature, frequently related by comparison or contrast to human behaviour. Rules relating to health and hygiene, desirable and undesirable foodstuffs, and prohibitions of incest and certain degrees of marriage, probably originated from generations of experience. Trial-and-error experience with agricultural and domestic practices must have occurred, though the biblical evidence for such activity is slight. Observation of nature is obviously present in the Scriptures. Experiment must have been present in a limited way, especially in matters such as agriculture and construction methods. Theoretical Explanation and Prediction are almost totally lacking, except to a limited degree in relation to the weather. The existence and action of God seem to have been sufficient explanation for the many unexplained phenomena of the natural world.

Beyond the limits of science: Supernatural beings, wonder and worship

An essential assumption behind this book is that scientific enquiry is a valid means of investigating and discovering reality, and that science may raise valid questions about religious claims.

At certain times in the history of human knowledge, information deriving from scientific investigations has caused profound rethinking of religious thought and language. Most notably, the heliocentric astronomers of the sixteenth and seventeenth centuries provoked a revolution in theological thought about the earth and its inhabitants and their place in the universe – in the twenty-first century we have become so accustomed to the new worldview that we find it difficult to appreciate the depth and seriousness of the changes brought about in Christian theology through the work of mathematicians and astronomical observers such as Copernicus and Galileo. In the nineteenth century, geological enquiries into the age of the earth, and the explanations offered by Charles Darwin for the diversification of species through natural processes, led to further rounds of soul-searching and theological re-writing, for Darwin himself and for many religious believers. In some religious communities, these challenges are still present. Just as it took a century and more for the astronomers' claims that the earth was spinning on its axis while it revolved around the sun to be generally accepted, and even longer for Galileo personally to be formally reinstated by some church authorities, in the same way the findings of science about the age of the earth, the origins of the universe and the evolution of living species have yet to penetrate some areas of Christian thought and theology. Twentieth century findings on matters such as relativity, quantum mechanics and genetics, which together question our traditional understanding of time and space, the meaning of reality in the physical world, the nature of living material and the relationships between species, have hardly begun to be regarded as necessary data for theological thought. While scientific research has been a growth industry, theological training has been in decline – a huge gap has opened, in need of urgent attention.

However, although scientific findings challenge theological claims about reality, we should not overestimate the potential of science to address religious questions. A part (though not the whole) of the function of religions is to meet human needs, and religions therefore address matters of concern to human life on the earth. Science may rightly question and even correct religious assumptions about issues such as the nature of the earth, its age and origins, the nature of disease, and others. But some aspects of religious belief are beyond the scope of science to address: the limits of science must be recognised, as well as its achievements.

Is it possible for science to contribute information relating to the existence or non-existence of God? Some scientists behave as though the question of the existence of God is capable of resolution, either positively or negatively, by scientific means, while some religious believers take it for granted that science cannot possibly contribute to this debate. In fact, the answer is bound up with the nature of the God in question. If God were a physical entity existing at a specific location in the universe, it might well be possible for such a God's existence to be demonstrated or dismissed by scientific means. If God communicated with people by physical channels such as wires or radio frequencies, again it might be possible for scientists to detect the messages and comment on their reliability. But the God believed in by most religious people, certainly by Christians, is not of this character at all. God is spirit, and the realm of reality described by the word spirit is, by definition, outside the physical order and therefore impossible to detect by scientific methods, though it interacts with human awareness at some level. The abilities of different people to detect this otherness, and the relationship of these abilities to cultural or genetic factors, or to specific experiences, are large and fascinating questions. Scientific enquiries may well elucidate the means by which people become aware of a religious dimension, but even so will not be able to pronounce on the reality of the phenomena described.

If the existence of God lies beyond the capabilities of science either to prove or to disprove, so also is the possible existence of spirit in other forms. The Bible contains many examples of interviews with beings described as "angels", in both the Hebrew Scriptures and in the New Testament. These beings are usually described in very human terms, in that they can hold conversations and share food with the human characters in the story – for example, the visitors to Abraham who enjoyed a quality meal (Genesis 18.8), or the soldier who was asked by Joshua whether he was "one of us, or one of our adversaries" and who identified himself as "commander of the army of the Lord" (Joshua 5.13-15). Joshua was told to remove his shoes as the place of the interview was holy ground, thus confirming Joshua as the successor of Moses. These visitors are always regarded as male – I am not aware of any female examples in the Bible. Only rarely are they described in the traditional terms, with shining robes and wings.[118] Sometimes they are alone, sometimes in groups.[119] Frequently the text contains ambiguities as to whether "the angel of the Lord" or "the Lord" himself is present.[120]

In interpreting these and other biblical accounts of the work of angels we must take account of the need which was felt for intermediaries between God and humans. The entire biblical faith is founded upon the view that God communicates with humans, but how does he do this? God communicates sometimes through inspired people, sometimes through prophetic messages and visions, sometimes through laws, sometimes through the liturgy of shared worship, sometimes through historical events, sometimes through Jesus Christ, and sometimes, according to the stories, through angels. Scientists, along with others, will have their own views about how to evaluate these various routes of divine communication, but it should not be supposed that the methods of science themselves can adjudicate on the matter.

One interesting aspect of the manner in which God communicates is the frequent reporting in the Bible of what God "said" or even "thought". Prophets and other holy people were so convinced that their inner thoughts and experiences came from God that they were ready to claim their own words as the word of the Lord. They were sufficiently certain and bold to say, "Hear the word of the Lord". For others the message came from mystical visions (a surprisingly infrequent phenomenon in the biblical story, but present in the books of Ezekiel, Daniel, Revelation and possibly some others of the prophets). In Genesis 15.1 we read that the word of the Lord "came to Abram in a vision", suggesting a less direct method of communication than the more usual "God said". On yet further occasions the biblical text lies plainly in the realm of storytelling. No one was present to hear what God "said" at the time of creation (Genesis 1.3, 6, 9, etc), or how God talked to no specified hearer when he commented on the wickedness of mankind (Genesis 6.3, 7) – the accounts were written much later than the purported events. We are even told what the Lord said "in his heart" (Genesis 8.21) or what "God thought" (Exodus 13.17). The language is that of a story about God and his relationships with his people.

Indirect evidence for the means by which God's voice was sometimes perceived may be found when we read that Rebekah "went to enquire of the Lord", implying a special place where the Lord's answers were found (Genesis 25.22). In this early incident the place to meet God was probably a small domestic shrine – we are told that Laban had household gods (Genesis 31.19). According to Exodus 25.21-22 and Numbers 7.89, it was from the space between the two cherubim above the ark that God spoke. The use of Urim and Thummim to discern the will of God has been discussed in Chapter 5.

Another example of mysterious phenomena beyond the scope of science either to explain or to explain away occurs in the references to "seeing God". Although there is a longstanding tradition in the biblical texts that it was impossible to see God or to hear his voice[121], there are occasions when individuals came perilously close, and said so. Some of these instances are associated with the encounters with "angels" examined above. Manoah was terrified that their vision of an angel guaranteed that he and his wife would die, but she sensibly pointed out that God would not have accepted their sacrifice nor informed them about the coming birth of a child if he intended to kill them (Judges 13.22-23). Moses was very close to seeing God at the incident of the burning bush (Exodus 3.6) and on Mount Sinai, where we are told that Moses and other leaders and seventy elders "saw the God of Israel" (Exodus 24.9-11), in contradiction to the general prohibition and without suffering any harm – on the contrary, "they beheld God and they ate and drank" (v11), presumably some form of a meal confirming a covenant. When Moses asked to see God's "glory", God proclaimed his name "the Lord" and revealed to Moses only his back (Exodus 33.18-23; cf. Numbers 12.6-8 and Deuteronomy 34.10 which maintain that Moses did indeed commune with God "face to face", literally "mouth to mouth" [122]).

Especially close encounters with God are frequently attributed to God's "spirit" or breath, seen as an expression of God's presence or an intermediary to avoid the danger and impossibility of a direct meeting with God. In the Hebrew Scriptures the spirit of God could provide skills for the construction of the tabernacle (Exodus 35.30-35), abnormal physical strength as with Samson and Saul (Judges 14.6, 19; 15.14; 1 Samuel 11.6), discernment and the requirements of leadership (Judges 3.10; 1 Samuel 16.13), military prowess (Judges 6.34), the declaration of verse (1 Chronicles 12.18), and prophecy (in many places, including 1 Samuel 10.6-13; 19.20-24; 2 Chronicles 15.1; 20.14; 24.20, Ezekiel 11.5).

In the New Testament the Spirit of God comes upon Jesus in a manner somewhat reminiscent of the earlier prophetic experiences (Matthew 3.16; Mark 1.10; Luke 3.22) and Jesus' own ministry is seen by Luke as directed by the Spirit (Luke 4.1, 14, 18; 10.21). After the resurrection of Jesus, the Spirit is seen as given to the whole group of Jesus' followers, whether through the presence of the risen Christ as presented by the gospel of John (20.22) or by the direct receipt of the Spirit in Luke's account of the day of Pentecost (Acts 2.1-4).

The development from these descriptions of the Spirit as the experience of the presence of God or of Christ into the full Trinitarian doctrine of the Christian church, in which the Spirit is regarded as a "Person" of the divine Trinity, is a story occupying several centuries in the patristic period. For our present purposes it is sufficient to note that spirit as a category is outside the reach of scientific analysis – spiritual issues are matters of Christian experience and subjects for theological consideration. They cannot be measured by scientific methods.

We have also passed the limits of scientific explanation when considering the practices of worship and praise to God. Worship is a response of the human being to the divine reality, and just as the reality of God lies outside the scope of scientific study to determine, so also the practice of worship cannot be subjected to scientific judgment. Practices of worship and prayer may of course be studied and compared sociologically, and the effects of worship on the human brain and neural systems may also be observed, but these investigations do not determine the reality or otherwise of the objects of worship. Many scientists express "awe" in face of the wonders of the universe; the worship of God extends this awe to the recognition of a loving being towards whom the human life may develop a relationship.

7

The choice: what should change and what remain?

The change to a scientific culture

A scientific outlook and a scientific mode of thought dominate our present western society. Even those people who do not regard themselves as scientifically trained or interested in scientific matters think in a scientific way: they consider that claims should be verified by evidence before they can be confidently accepted, and that events which affect our lives usually happen because of preceding events. These are major parts of the mental attitude I am describing as scientific. The fact that most people will regard these attitudes as normal, or self-evident, or common sense, indicates how deeply rooted the empirical approach has become in our mentality. It was not always so.

In earlier times and cultures, including the times and cultures represented in the biblical accounts, there existed a much greater awareness of a non-physical, non-scientific level of existence which could affect events in this world. This other level might be described in terms of the Jewish and Christian understanding of God, or some other god or gods, or angels, or spirits and demons both beneficent and malevolent, or astrological powers in the heavens, or magical powers possessed by special persons. Whatever its form and means of expression, this other world had a profound influence upon this world.

Scientific knowledge and explanations of phenomena have contributed to the steady weakening of this understanding of the influence of external powers – what has been described as the "disenchantment" of the world.[123] Many events in the world previously attributed to divine action are now prosaically but convincingly explained in terms of physical or biological processes. This applies not only to awe-inspiring though everyday events such as the appearance of a rainbow, or a thunderstorm, or a lunar eclipse, or the birth of a child, the reasons for which are now understood by even very young children, but also to less visible happenings such as the progress of a disease or the causes of psychological illness.

In recent decades there has been in some societies a reaction against a solely material approach to explanations. This may be seen, for example, in the growth of practices such as meditation and what may loosely be called "eastern" religious views in western societies, and also in the greater recognition of the role of religion in promoting health and healing. In my view it is important to recognise that this reaction against an entirely materialistic interpretation of the world does not negate the considerable explanatory power of the sciences, and the benefits thereby brought to human society. While some commentators set themselves in opposition to the role of science, very few are willing to abandon its achievements in, for example, the provision of safe electrical power, the relative ease of travel and communication, and the availability of healthcare.

This deep change in our mental processes brought about by scientific methodology raises important questions in those people for whom a religious or spiritual dimension is an essential component of their lives, and particularly in those who regard highly the teachings of the Bible as formative in their religious outlook. For, as we have seen throughout this book, the cultural and mental assumptions which underlie the biblical text are in many ways different from our own. The biblical message is therefore difficult for contemporary people to absorb, and certain mental adjustments are necessary.

Where should these mental adjustments be made? Is it necessary for a twenty-first century reader of the Bible to abandon all the knowledge and outlooks accumulated over the centuries and enter entirely into the languages and cultural worlds of the biblical era in order to comprehend its teaching? Should our mental frameworks be modified in this way by the Bible writers? Alternatively, should the Bible story be "corrected" or reinterpreted in accordance with our contemporary outlook? If changes are necessary in both directions, where should the changes lie, and how should this be achieved?

The biblical view of divine action

To understand completely what was in an author's mind, including the emotions and unwritten presuppositions behind the text, is impossible. Some modern commentators therefore abandon the attempt and regard the imposition of their own assumptions or any other external view as a valid hermeneutical method. My own approach is to attempt to understand as completely as possible the author's sense and intention. To develop even a partial understanding of the biblical text requires some knowledge of the events and cultures contemporary with the original author and the text – centuries of theological and literary commentary by experts on the text have aimed to elucidate its meaning for other readers. With more information about the contemporary culture we will have a better understanding of the Bible, but there is clearly a limit to the extent to which most people will be able to take part in this process.

Moreover, *knowing* certain information about the culture of biblical times is not the same as *adopting* those views. Our central difficulty therefore remains – better knowledge of the cultural background of the biblical writers may help me to understand what they intended, but if I do not share their cultural background and some of their beliefs, how am I to use the biblical text in the formation of my own belief system?

The first step is to summarise what we have discovered about the biblical view of the natural world. Having established as best we are able the views, possibly diverse, of the biblical writers, we will then go on to consider the relationship between those views and scientifically based views, and the extent to which the biblical views remain credible or ethically acceptable.

The primary characteristic revealed through our examination of so much biblical material is the belief that *God was directly active in the natural world, and in the world of human events*. But exactly what did the authors mean by these claims? Four alternatives are possible.

First, it might be cynically suggested that the language is entirely *conventional* in attributing events to God. Conventional religious language is common in some circles, expressing, for example, thanks to God for certain events or promises of prayer for people in distressing situations, without any serious consideration of the implications of such an attribution, nor of possible alternative explanations. Conventional language about God may be used by politicians or authors aiming to please their audiences, regardless of their personal opinions. Arguably it was this approach used by Charles Darwin when in spite of his increasing disillusionment with established religion he wrote of a "creator". This interpretation is highly improbable for the biblical texts – perhaps it is a reflection on our critical society that this explanation even comes to mind. In several of the examples quoted in the preceding chapters the emphasis seems to be strongly on the reality of divine action. The prevailing belief of the biblical times in the predominance and power of God make it very unlikely that empty, conventional language would be used. Even those people who did not like what the prophets were telling them about God nevertheless believed in God's reality and importance. This possible explanation, that the biblical language about divine action is purely conventional, should therefore be dismissed.

Second, we might say that these biblical claims are related to specific, single events in which God was at work for his specific purposes, but these specific events should not be used to infer general principles of divine action. This might be called the *isolationist* method of interpretation: each biblical example is an isolated specific case, from which wider conclusions about the methods of God's operation in the world should not be drawn. On this interpretation, God was active in specific cases of child-birth or the failure to produce children, but is not necessarily active in every such case. God was active in the romantic feelings of Samson towards the Philistine woman, but is not necessarily present in all romantic feelings. God sent a deep sleep upon Saul and his troops in one set of circumstances, but is not necessarily present in every similar example. God gave wisdom to Solomon, but is not to be thought of as the source of all wisdom wherever it is found, nor the reason behind its absence in other cases. God is a protector or deliverer at times, but not always.

It is certainly the case that many of the biblical references to divine acts relate to single instances, without drawing any conclusions about divine action in other cases. For example, many of the accounts of individual deaths that are attributed directly to God do not go on to claim that God acts in this way in every case, and many of the accounts of God's judgement through disasters or wars do not attribute every similar event to God's direct will and intention. It may be that some of the authors would have shied away from this conclusion – but that is an argument from their silence, and therefore unreliable. It is possible that the biblical authors thought differently about these issues, or did not think about them at all.

We must therefore move on to a third method of interpretation which comes close to being the opposite of the isolationist method. In this third scheme, the specific examples of God's action described in scripture may be used, indeed should be used, to establish general patterns of divine action everywhere. This could be described as a *universalist* method of interpretation, as it is affirming that God is immanently present in all human experiences.

As we have seen in previous chapters, there are biblical examples of this universalist view of the activity of God in the world of humans and of nature. For example:

Who gives speech to mortals? Who makes them mute or deaf, seeing or blind? Is it not I, the Lord?
Exodus 4.11

See now that I, even I, am he; there is no god besides me.
I kill and make alive; I wound and I heal;
And no one can deliver from my hand.
Deuteronomy 32.39

The Lord kills and brings to life; he brings down to Sheol and raises up.
The Lord makes poor and makes rich; he brings low, he also exalts.
1 Samuel 2.6-7

It is God who executes judgement, putting down one and lifting up another.
Psalm 75.7

Whatever the Lord pleases he does, in heaven and on earth, in the seas and in all deeps. He it is who makes the clouds rise at the end of the earth; he makes lightnings for the rain and brings out the wind from his storehouses.
Psalm 135.6-7

He covers the heavens with clouds, prepares rain for the earth, makes grass grow on the hills.
Psalm 147.8

There is no one besides me; I am the Lord, and there is no other. I form light and create darkness, I make weal and create woe; I the Lord can do all these things.
Isaiah 45.6-7

The Lord our God, who gives the rain in its season, the autumn rain and the spring rain, and keeps for us the weeks appointed for harvest.
Jeremiah 5.24

Who can command and have it done, if the Lord has not ordained it? Is it not from the mouth of the Most High that good and bad come?
Lamentations 3.37-38

Does disaster befall a city, unless the Lord has done it?
Amos 3.6

[Your Father in heaven] makes his sun rise on the evil and on the good, and sends rain on the righteous and on the unrighteous.
Matthew 5.45

He has mercy on whomsoever he chooses, and he hardens the heart of whomsoever he chooses.
Romans 9.18

On this basis, it would be claimed, there is reason for affirming that God is active in every situation, both good and bad: in every conception, every childbirth, every example of romantic love, every case of human wisdom (and therefore, since wisdom is a relative commodity, in every example of the lack of wisdom), every instance of prosperity, and all cases of healing and protection (and therefore, again, for all cases in which prosperity or healing or protection do not occur). If God "put to death" Er and Onan, the two sons of Judah, God is also responsible not only for every sudden death, but also for every single death. In this extreme form the universalist model approaches a view of complete divine sovereignty, in which God is seen not merely as present in everything, but is understood to be responsible for everything, whether good or bad. There seems little doubt that some of the biblical authors held this view.

A fourth method of interpretation might be described as *theological* – the authors wanted to ascribe the outcomes to the purposes of God. They regarded any immediate physical causes as trivial in comparison with the action of God revealed in the event. They were not concerned with wider questions such as whether God was similarly active in all events of this type, or only in this event. Rather, they were telling the story of this event from a theological perspective, without reference to what happens in other events. This theological method might overlap with either the isolationist or universalist views described above.

It is possible, of course, that more than one of these methods of interpretation is present in the scriptural examples. One writer might have been describing God's action in what I have described as an isolationist manner, as defined above, while another might be universalist, and another theological. All these methods of interpretation are theological at some level, and the above categories are not entirely watertight. Individual scriptures need to be interpreted in their context, and a single overall conclusion is not possible. In general those parts of the Scriptures which are dealing with non-specific situations, such as the wisdom literature and many of the Psalms, are more likely to be offering an interpretation of God's action related to a wide range of situations rather than merely to individual events.

Parallel explanations

In these chapters I have shown the common biblical theme that God is intimately active in the natural world, controlling natural events such as earthquakes, droughts and plagues, and human events such as conception and childbirth, the rise and fall of nations, health and illness, and death. A further question arises: does this understanding of divine action allow for other parallel causes, which might be more consistent with scientific explanations?

We have seen that in many instances a role for physical causes is recognised, even though the theological interest of the writer placed the emphasis on the activity of God. Most obviously, the importance of clouds in the production of rain, and to a lesser extent the significance of wind direction, are recognised, even though God is described as the source of rainfall. The regularity of seasonal variations is also recognised, along with the importance of agricultural skills derived from generations of experience, even though God is regarded as the giver of good harvests. While some deaths are attributed to God, the function of disease is also recognised, and measures taken to combat it where possible.

It thus appears that at least some of the biblical authors were able to accommodate parallel explanations of natural phenomena: God is regarded as the real reason for an event, though other factors are recognised. We might be tempted to say that God was the "ultimate cause", but such language tends to place God in a background position, whereas the biblical language more frequently places God in the foreground – more like a primary cause. But the traditional theological language of primary and secondary causes has its own difficulties. How do the two levels of cause relate to each other? Does one overrule the other, and how, and when?

It is possible that some of the biblical writers really believed that God was directly and immediately responsible for all events in the natural world. This comes close to an animistic type of religion, in which God is present and active in every aspect of the physical world. Certainly the God presented in parts of the Hebrew Scriptures can appear unpredictable, violent and arbitrary. This picture of God seems very different from the loving and forgiving Father presented in some of Jesus' parables, or the God of grace described in some of Paul's letters, or a God of universal love in whom twenty-first century Christians, weary of violence, would prefer to believe – or, indeed, from the good shepherd who makes me lie down in green pastures.

A more probable approach would be to invoke the ancient ignorance of the true causes of events. People did not understand why rainfall in successive years could be so different, why an earthquake could strike out of nowhere, why some couples could produce offspring and others were unable, or why apparently healthy people should drop dead without warning. Given this uncertainty and unpredictability, against the background of a faith in Yahweh the Creator, it would be understandable to resort to divine activity in these unexplainable circumstances. Such a view comes close to fatalism – nothing can be done about it, we have to accept the divine will. More facetiously, and in a different culture, it might be described with a shrug of the shoulders as "the will of the gods".

As noted above in the discussion about the thinking behind the scriptural writings, a further option is to regard these biblical descriptions of divine action as expressions of theological selectivity. Simply put, it suited the view of the world held by the historian, prophet or psalmist to attribute particular events to God's action. The sudden death of a wicked person or the appearance of a swarm of locusts could be explained as a divine judgment, given the state of the individual or nation at that time. This did not mean that every sudden death or locust attack should be explained in this way, or that all wicked persons would meet the same fate. The inconsistency of this form of theological selectivity is one of its biggest weaknesses – if not all events of this type are to be explained in this way, why should we accept a theological interpretation of this event? But such objections did not seem to dampen the prophetic inclination to read disasters in this way.

In summary, it may be helpful to distinguish between two distinct processes – of *explanation* and of *interpretation*. Although frequently confused, these processes are not the same. The scientific approach is to search for explanation, in describing the physical phenomena that have caused particular outcomes. This is essentially a backward-looking process, though if it is done well it can lead to successful predictions. The theological approach is towards interpretation, enquiring about the purposes and possible reasons for events. This is, of course, a more speculative and subjective activity, open to alternative views and rejections. Scientific explanations often command widespread acceptance, especially when they have been verified many times with cumulative evidence. Theological interpretations will always invite personal reactions in the form of faith – yes, I accept that this is a true expression of why something happened, or no, I do not accept that interpretation. A part of our difficulty in grappling with biblical ideas if that we are constantly looking for explanations while what we are being offered are interpretations.

Is it credible?

We must now face the question of the extent to which the biblical views we have been exploring are acceptable in our present times. This question has to be addressed at three levels: scientific, ethical and theological. Given the many ways in which science has clarified the operations of the material world, how much of the scriptural view of God's activity can be retained? And given the emergence and development of Christian ethical principles, is it necessary to regard some theology in the Bible as of merely historical interest? And are some of the theological foundations of the Bible still tenable?

First we address the scientific question. Undoubtedly the progress of scientific knowledge during the last four hundred years or so has revealed to us truths about the natural world that are in conflict with the biblical worldview. Our understanding of the solar system, the age of the earth and the universe, and the causes of illness and death mean that many biblical statements can be read only, at best, as poetical. This does not deny the beauty and the seriousness of the biblical writings – they were produced within the limits of contemporary assumptions. But we have to conclude that the Hebrew cosmology was wrong, that beliefs about the origins of humans and the significance of gender were wrong, and that ideas about the permanence of the earth and the stability of coastlines were wrong. In recognising these limitations to the biblical view of the physical world, we do not diminish the religious teachings of the Bible.

Turning to the ethical question, it must be remembered that Jesus Christ raised the ethical requirements based on the Hebrew Scriptures, both explicitly ("you have heard that it was said ... but I say", Matthew 5.21ff) and implicitly: his statement "let the dead bury their own dead" (Matthew 8.22, Luke 9.60) was a profound rejection of many of the principles of Hebrew faith as practised in his times. It follows that practices endorsed in the times of the ancient Hebrews are up for reconsideration. Moreover, since the times of the New Testament numerous practices which were then acceptable, such as slavery and the social treatment of women, have been subjected to further examination and rejected. So, again, the biblical requirements have to be opened to fresh examination in the light of our present-day knowledge and conditions.

At the theological level, the idea that God is directly responsible for every natural event begins to look theologically unacceptable. If pressed to extremes in such a way, this view would lead to a world in which not only humans were mere robots acting out the divine will, but also every creature, raindrop and breath of wind were fulfilling in clockwork fashion what God had directed. We must look for a more satisfying understanding. Discoveries about the physical causes of events such as storms and rainfall, earthquakes, sudden deaths and insect plagues, and the human causes of wars and geopolitical changes, offer a more accurate picture. But if God is displaced from God's role as the primary cause, is there any role at all? Is the idea of parallel causes, accepted by some at least of the biblical writers, at all viable?

I have argued that the existence of God lies beyond the ability of scientific methods to either prove or disprove. The human mind will tend to locate God wherever other explanations are not available. For the ancient Hebrews, the circle of knowledge of the world and its operations was severely limited, and so God was frequently invoked as the cause of the unknown. As the circle of knowledge expanded, the references to God as an explanatory cause shifted their ground; God became responsible for that which remained unexplainable and, in the views of some, remains in that position. The danger of this view is, of course, that God will continue to retreat as human knowledge answers more and more of the outstanding questions.

The idea of God as a parallel cause is rather more subtle. Scientific and other explanations of events are not denied, but faith retains God as one involved in the process. This approach prevents the displacement of God too remotely, at the cost of introducing an uncomfortable second level of causality. God must be seen as a "cause" in a sense totally different from physical or human factors. God does not exert forces or push buttons or pull strings, except in a metaphorical sense. It is a matter of personal belief whether this additional explanation is accepted.

Is the Bible inspired?

The Bible displays several different forms of inspiration. One of the most dramatic is the report of King Saul's encounter with a group of prophets when "the spirit of God possessed him" and he fell into a "prophetic frenzy" (1 Samuel 10.9-12). A similar incident is described at 1 Samuel 19.20-24, when Saul's messengers also experienced "prophetic frenzy" and Saul himself stripped off his clothes and lay naked all day and night. These events were apparently the basis of the saying in ancient Israel "Is Saul also among the prophets?" meaning that people sometimes behave in an inconsistent and irrational manner. A possibly related factor is Saul's tendency to depression and violence, for example when "an evil spirit from God rushed upon Saul and he raved within his house" and he attempted to murder his rival (1 Samuel 18.10; 19.9-10).

So here is one understanding of the nature of prophetic inspiration: irrational, frenzied behaviour, including some possibly undesirable actions, attributed to an onrush of divine spirit, sometimes good and sometimes evil.

But within the Bible this is an exceptional form of inspiration. Only a few of the biblical texts display characteristics of a visionary or ecstatic kind, and even these do not indicate the excesses of Saul and his band of prophets, nor do they remove the prophet from the historical and cultural situation. Only parts of the prophecies of Ezekiel, some of the "Thus says the Lord" oracles of other prophets, and the Revelation to John are of this visionary character. The gospels describe a small number of visionary moments in the experience of Jesus, as at his baptism (Matthew 3.16-17), his transfiguration (Luke 9.28-36), in some healings (Mark 7.33-34) and a few other times (Matthew 11.25-27; Luke 10.18-22; John 3.13).

Most of the biblical texts are more prosaic: histories of Israel and lives of Jesus written from a particular dynastic or theological point of view, letters to young and growing churches, legal codes. Even the poetry and wisdom writing of the Psalms, Proverbs, Job and the songs distributed through the other texts are mostly of a carefully constructed rather than an ecstatic character. Some are even acrostics based on the Hebrew alphabet, such as Psalms 9, 10 and 25, and others appear to be parts of worship liturgies, such as the hymn to Christ in Philippians 2.6-11. These are not spontaneous outpourings of spirit-possession, but meticulously worked pieces of artistry intended for the benefit of readers, hearers or worshippers.

This variety in the experiences of the biblical characters and biblical writers should lead us to expect a similar variety in the nature of their inspiration. A single model of inspiration is not likely to apply to a set of texts as varied as those within the Bible.

In addition to recognising these various literary genres, we should also note the range of linguistic, historical and political contexts which have shaped the books within the Bible. Three languages are present, Hebrew, Greek and Aramaic, and those languages carry internal meanings which can never be fully recaptured even in the best translations. The influence of prevailing historical, economic and political circumstances is particularly evident in the works of the Old Testament prophets – would there have been a Jeremiah without the Babylonian threat to Jerusalem, or an Ezekiel without the exile? The gospel accounts of the life of Jesus clearly display differences according to the nature of the Christian communities within which they originated or for which they were intended. The letters of the apostle Paul were written to address the joys and problems of groups of believers in particular cities around the Mediterranean. Even Paul would have been shocked to be told that he was creating "holy scripture".

So, given this highly variable nature of the texts contained within the Bible, in what way are they to be regarded as inspired? Some of the ways in which inspiration have been understood do not take adequately into account the human and cultural situations in which the texts came into existence.

According to the theory of verbal inspiration by God the words of Scripture become, in effect, the very words of God. This has for long been the understanding of the more conservative parts of the Christian church. Although most of those holding this view also recognise, in theory, the cultural background and personal characters of the human authors, this context is usually overwhelmed by the view of divine authorship, leading to a supremely reliable and even inerrant or infallible text. However, this understanding of inspiration faces two particular difficulties.

First, the text includes numerous real though remarkably limited inaccuracies and contradictions within the text.[124] Inconsistencies of this kind, while possibly strengthening the genuineness of incidents reported by several witnesses, nevertheless are fatal for theories of verbal inspiration by God. Second, the Bible raises ethical problems related to practices once regarded as acceptable but now rejected. These include the use of the death penalty for certain offences that would not be treated in that way by later societies, the instructions by God to perform violent actions now regarded as unacceptable, and the low value and poor treatment of females. While these practices can be explained in terms of the assumptions and expectations of the times, ethical difficulties arise if the biblical statements that God ordered these things are taken literally.

The defence of verbal inspiration against these objections sometimes includes recourse to the "original autographs". These were supposedly the actual texts inspired by God, any minor errors creeping in through subsequent copying and transmission. The possibility of such corruption of the text is plausible and may resolve some of the textual problems, but it does not resolve all of them nor the ethical problems, neither does it recognise the process of editing, perhaps repeated several times, that has occurred in the transmission of many of the biblical books. Moreover, the "original autographs" remain unknown, so the theory does not provide the comfort and reliability that its proponents suppose. Recourse to original autographs is a counsel of despair.

Views of the inspiration of the Bible sometimes make use of the principle of accommodation, supposing that God was accommodating himself to the views of the times. John Calvin likened God's Word to the "prattling" of a nurse to a baby – God was working with the limited capabilities and understanding of ancient peoples. But does this theory of accommodation really let God off the hook of dubious moral instructions? Presumably, under the accommodation theory, God himself held to higher ethical standards, and knew that the standards would change over time even if the historical actors did not. Therefore God remains under the charge of ordering an action that was contrary to higher ethical principles. Is it not easier to conclude that when the Bible says in certain places "God said" do such-and-such, what it really means is that *someone believed* that God said do such-and such, or that the account is part of a narrative that has to be read rather more freely?

For many Christians the views and practice of Jesus must be definitive for all circumstances. The gospels inform us that Jesus quoted Scripture in his defence against temptation, but using the text as a source of strength in the spiritual journey is not the same as holding to a belief in its divine, faultless nature. It is also reported that Jesus used the Scriptures to explain to the disciples that the Messiah had to suffer and to rise on the third day (Luke 24.44-46). But again, there are many ways that images in the Old Testament can be used to illustrate events in the New while stopping short of a view of inerrant inspiration. The nearest Jesus seems to come to making a general statement about Scripture is at John 10.35, "the scripture cannot be annulled" (NRSV). But this comes in the middle of an argument with opponents about the meaning of particular words in Scripture: "*If* this, and *if* that, then ... But *if* something else, then ..." Jesus is here arguing on their terms, not necessarily stating his own view.

But even if this was Jesus' own view (and there is no reason to disbelieve that he held a high view of Scripture such as was prevalent among Jews of his day) was it equivalent to the view of inerrancy held by many today? The New Testament writers themselves displayed a remarkably varied attitude in their "quotations" from the Old Testament. In one place Paul bases his argument on the use of a singular rather than a plural noun, even though the word in question, "seed", could be regarded as a collective noun (Galatians 3.16 – an example not so much of verbal inspiration as inspiration of the letter!), and in Matthew's very Jewish gospel Jesus shows respect even for the punctuation marks of Scripture (Matthew 5.18). But at other times the New Testament writers did not respect the letter or the word of any version known to us, and some of their supposed quotations are difficult even to identify. They may have been expressing their understanding of a Scripture passage, but they were not verbal literalists. It is worth noting that while the Christian church has always given a high place to the Bible in guiding its beliefs and practices, the more extreme views of its inspiration emerged in the nineteenth century, in response to both the hyper-critical treatment of the text in some Protestant theological circles, and the promulgation of papal infallibility in the Roman Catholic church.

If these views of inspiration discussed above are to be rejected, what form would a more moderate version of inspiration take? I propose that the following principles should be part of a view of the inspiration of the Bible that takes account of its historical origins, including its limited view of scientific realities:

- Different parts of the text will display differing forms of inspiration. A single model of inspiration, imposed over the whole text, will not fit the large variety of authors, cultures and circumstances prevailing during the centuries of the text's creation.
- These human circumstances and historical settings of the biblical writings must be recognised to the fullest extent possible. Our appreciation of the text will be enhanced by the greatest possible knowledge of its original sense. Where our knowledge of the world based on scientific findings differs from the worldview of biblical times, this must be recognised and incorporated into our understanding of the Bible.

- A direct form of inspiration by God may exist in a small number of biblical texts, such as some of the prophetic oracles ("This is the word of the Lord") and in some visions (Ezekiel, Revelation), though even in these cases the cultural context of the words is apparent. Inspiration has not removed the writer into a different world, but direct inspiration by God is being claimed and may be recognised.
- Most of the Bible was written as history, law, poetry, wise sayings, gospels, instructions. The authors do not claim direct inspiration by God and did not think they were writing "scripture". In some cases they must have known that they were revising previous versions (for example, Chronicles' rewriting of Samuel/Kings, prophetic revisions of earlier prophets, Matthew's and Luke's use of Mark, and the work of editors generally). Therefore inspiration has to be understood in a looser sense, similar to that in which poetry or music or a sermon or a conversation may also be said to be inspired, drawing from a source which lifts and transforms the reader or hearer. Inspiration occurs within the human experience rather than being introduced from outside by God.
- Nevertheless, Christians would wish to affirm that the Bible is not merely a human book. God was working in a revelatory way even though some of the beliefs and practices involved became unacceptable in later times. I propose that the inspiration of the Bible should be thought of as a *Contextual Response to God*. The influence of the divine is present, though it is limited by the knowledge and circumstances of the recipient. This is not the same as the "accommodation" theory criticised above, in which God accommodates himself to the ideas of the people, including to their ethical ideas. In the system I am proposing, the limitations are definitely on the human side.

- Ethical principles arising from the teachings of Jesus and the faith of the New Testament have to be applied to the earlier texts, leading to the rejection of some of the legal, ethical and ritual requirements of the Hebrew Scriptures.
- Even the teachings and practices of the New Testament have to be subjected to a similar examination using ethical standards based on the enlarging application of Christian principles. In the past this has led to a revision of views commonly held concerning slavery, and today it is relevant to a new understanding of gender differences and possibly other topics also.

Do these principles to be incorporated in our understanding of inspiration lead to a less absolute, more risky, use of the Bible in Christian discipleship? Yes, and they should be welcomed also for that reason. The trust of Christians should be placed in the living God, not in the words of a culturally conditioned book, even though the book reveals God to us. As John Polkinghorne has expressed it, we should not be afraid of "intellectual precariousness" or "cautious daring in the quest for truth".[125]

Concluding remarks on Biblical and scientific views

The question addressed throughout this book has been 'How can scientific knowledge and the belief systems propounded in the Bible be brought together?'

The knowledge achieved by several centuries of scientific investigation cannot be ignored, forgotten or un-discovered. There is no going backwards to a time when the physical structure of the universe was unknown, before the earth's formative processes had been elucidated, before the DNA shared among all living things was revealed. The place of scientific knowledge in our own mental frameworks has to be recognised, celebrated and preserved. Christians cannot produce their own versions of science to ease the cultural divide between the pre-scientific and the scientific ages – that would be too easy an escape from the basic problem, namely that Christianity relies upon documentary material which arose in very different cultures two to three thousand years ago.

This is not to say that the scientific community is always "right" in the views it holds about the phenomena or processes it investigates. The history of science contains many revisions and not a few reversals of the opinions prevailing at any given time. But these revisions were not arbitrary, invented to fit with a religious understanding of the world. Rather, they were based on further physical evidence and/or the solution of awkward problems that remained in the previous explanation. Scientific discovery is a moving process which builds upon, revises and sometimes rejects previously held views. Science is also a process which relies upon the broad acceptance within its community of any given interpretation or explanation. Rival theories may, and occasionally do, overthrow the prevailing consensus, but this does not permit religious believers to reject scientific views for religious reasons.

Believers may sometimes hope that scientific advances themselves will introduce changes which move scientific opinion in a more religiously acceptable direction. For example, the generally recognised success of the Big Bang model of the beginning of the universe over and against the steady state model, which seemed to be a serious rival around the middle of the twentieth century, was a change which many believers welcomed as support for belief in creation and a Creator.[126] However, believers should also recognise that the results of scientific research may also have the opposite effect, rendering certain religious views harder to maintain in the face of scientific evidence. One area where this may be happening at present is in relation to the human brain and the human "person". One religious interpretation of the human person is that whatever may be known about the functioning of the human body and the human brain as a biological organ, there remains a "person" or "soul" with needs which can be fully addressed only from a religious perspective. At some point in the future this may become an argument facing increasing difficulty, as the nature of the human "person" is clarified. A primary principle, therefore, in the relationship between Christian belief and science is that the Bible reader has to be open not only to scientific knowledge from the past and present, but also to the possibility of relevant scientific knowledge in the future.

A further important principle in developing a manner of using the Bible which is consistent with scientific knowledge is that the biblical texts have to be read to the fullest extent possible in their historical and cultural context. This should be abundantly clear from the preceding chapters of this book. The biblical texts did not come from heaven on a cloud nor were they (for the most part) revealed in mystical visions. They were addressed primarily to specific situations existing in the nation of Israel or the first-century Christian church, and have to be understood in this light. Even without the questions raised by modern science, a process of reinterpretation exists in the transfer of the text from the original languages into, for example, the English language, and from the first century CE or earlier into the present. Scientific questions merely represent a further stage in the process of cross-cultural transfer already begun.

From our point of view within the immediate purposes of this book, this recognition of the cultural origins of the Scriptures will take account of the non-scientific nature of the societies in which the texts originated. Although some knowledge of natural processes was present, explanations of events were as likely to be found in divine activity as in what we would now describe as scientific processes. As we have seen, this means that some diseases were regarded as divine punishments, some wars were described in terms of God's work on behalf of his people or God's judgements on sinfulness, and some unexplainable events were reported as divinely enacted miracles. Because scientifically rigorous explanations were absent, these descriptions were prone to exaggeration. And because scientific awareness was as absent in the receivers of these accounts as in the sources, the accounts became an accepted part of the Christian church's religious heritage.

These two principles – that our scientific knowledge has to be maintained and respected, and that the cultural origins of the biblical text have to be recognised – may seem to be no more than a restatement of the question with which this study began: namely, how should a scientifically oriented person approach the religious material in the Bible? Can we do no more than hold in tension the diverse cultures of the biblical world and the modern world, and try to find a personally acceptable route between them? A few further principles will help us to handle this problem.

In spite of our respect for the many achievements of science, it is also necessary to recognise the limited abilities of the sciences in handling religious matters. In describing physical and biological phenomena the sciences have added enormously to our knowledge. On the whole, this knowledge has been used beneficially for human life, though with some notable exceptions. But in fields of human behaviour and human relationships, the achievements of science have been much more limited, though as I have emphasised above, it would be wrong to assume that this will always be the case. Most importantly, it is necessary to recognise that there are areas of human experience and perception in which science has no role at all. God is, of necessity, beyond the area in which measurement and proof are available. Experience of God can be used as evidence, though it will always be capable of alternative explanation and will fall short of secure proof. God is outside the circle in which scientific explanations can operate. And if that is true of the being of God, it may also be true of experiences and events associated with the being of God, such as the reality of worship and other experiences or beings at a spiritual level.

It is also necessary to establish priorities in the values we place on the various sections and books of the Bible. Old-fashioned preachers may exhort us to believe that all parts of the Bible are inspired, are of equal worth, and exist for our benefit, but the reality is that readers of the Bible, and preachers too, will exercise judgement in deciding which parts of the text are of greater and lesser value. For most people, the teachings of Jesus are of greater worth than the laws about sacrificial practices or clean and unclean animals, and the doctrinal writings of the apostle Paul more intellectually demanding and ultimately satisfying than the accounts of his travels. This is not to deny that all parts of the Bible are valuable in giving us insights into the beliefs and religious practices of Hebrew and early Christian times – as I hope to have shown in earlier chapters of this book, even the Levitical laws yield valuable information about how people thought about the world and about God and the proper way to worship him. In earlier times expositors used allegorical methods of interpretation to squeeze some value from the more obscure parts of the text, but such approaches are no longer regarded as acceptable, and the alternative is to regard some parts of the Scriptures as of lasting worth, and other parts more firmly attached to the times of their origin. There are, of course, not merely two categories, but a sliding scale with an infinite range of values – the most precious parts will include some culturally determined words or ideas that we may not grasp, and those texts which appear totally related to events far away and long ago will nevertheless offer valuable insights. Where each reader locates texts on this sliding scale will be a matter for individual assessment.

This subjectivity is anathema to some Christians, particularly those who wish to impose uniformity on Christian belief and practice. But I believe it is a necessary part of our reading of the

Scriptures and relating what we read to our very different scientifically knowledgeable world. Moreover, we need to remember that the biblical writers themselves exercised subjectivity in their use of preceding texts. Prophets re-used and re-applied the words of earlier prophets, while the New Testament writers used the version of the Hebrew Scriptures that most closely suited their theological purposes, and sometimes paraphrased or combined texts to make their required point. Selectivity among texts has strong precedent within the Bible itself.

A further consideration lies in the area of ethics and moral sensibilities. Many of the changes between the biblical worlds and our own occur in matters of ethical or moral behaviour. And where we see an improvement, it is right to reject the earlier version. So we abhor the nationalistic fervour which underlies much of the worship of God in the Hebrew Scriptures, and prefer a theology in which all people are equally loved and valued. We believe we have progressed beyond the attitudes of male superiority which exist in both the Jewish and the Christian Scriptures. We wince at some of the punishments meted out for crimes which we would recognise as matters of personal choice. Our sense of ethical value therefore guides us, beyond any considerations of scientific evidence, towards those parts of the Scriptures of greater value in our own times.

To summarise my findings about the relationship of science to the use of the Bible, we should maintain and celebrate the findings of science while also recognising its limitations in dealing with religious issues; we should learn as much as possible about the history and culture of biblical times, which will lead to a degree of selectivity in the value placed on various books and sections; and we should apply Christian ethical principles in determining those parts of the Scriptures of greatest value.

Bibliography

W. F. Albright and C. S. Mann, *Matthew*, The Anchor Bible, Doubleday, New York, 1971

Leslie C. Allen, *The Books of Joel, Obadiah, Jonah and Micah*, Eerdmans, Grand Rapids, 1976

Timothy R. Ashley, *The Book of Numbers*, Eerdmans, Grand Rapids, 1993

Richard J. Bauckham, *Jude, 2 Peter*, Word Biblical Commentary, Volume 50, Waco, 1983

G. K. Beale, *The Book of Revelation*, New International Greek Commentary, Eerdmans, Grand Rapids, 1999

Joseph Blenkinsopp, *Isaiah 1-39*, The Anchor Bible, Doubleday, New York, 2000

Joseph Blenkinsopp, *Isaiah 40-55*, The Anchor Bible, Doubleday, New York, 2002

Joseph Blenkinsopp, *Isaiah 56-66*, The Anchor Bible, Doubleday, New York, 2003

John Hedley Brooke and Ronald L. Numbers, *Science and Religion around the World*, Oxford University Press, 2011

W. Brueggemann, *The Message of the Psalms*, Augsburg Old Testament Studies, Minneapolis, 1984

Philip J. Budd, *Numbers*, Word Biblical Commentary, Volume 5, Waco, 1984

Brevard S. Childs, *Isaiah*, Westminster John Knox Press, Louisville, 2001

Duane L. Christensen, *Deuteronomy 21.10 – 34.12*, Word Biblical Commentary, Volume 6B, Thomas Nelson, Nashville, 2002

Francis Collins, *The Language of God*, Pocket Books, Simon and Schuster UK Ltd, London, 2007

C. E. B. Cranfield, *Romans: A Shorter Commentary*, T&T Clark, Edinburgh, 2001 (1985)

Oscar Cullmann, *Christ and Time*, SCM Press, London, 1962

Mitchell Dahood, *Psalms III (101-150)*, Anchor Bible, Doubleday, New York, 1970

John Day, *Yahweh and the Gods and Goddesses of Canaan*, Sheffield Academic Press, 2000

Simon J. DeVries, *1 Kings*, Word Biblical Commentary, Volume 12, Waco, 1985

James D. G. Dunn, *The Acts of the Apostles*, Epworth Press, Peterborough, 1996

John I. Durham, *Exodus*, Word Biblical Commentary, Volume 3, Waco, 1987

Gordon D. Fee, *The First Epistle to the Corinthians*, New International Commentary on the New Testament, Eerdmans, Grand Rapids, 1987

Yehuda Feliks, *Nature and Man in the Bible*, Soncino Press, London, 1981

Michael V. Fox, *Proverbs 1-9*, The Anchor Bible, Doubleday, New York, 2000

R. T. France, *The Gospel of Matthew*, Eerdmans, Grand Rapids, 2007

Duane Garrett and Paul R. House, *Song of Songs/Lamentations*, Word Biblical Commentary, Volume 23B, Thomas Nelson, Nashville, 2004

Joel B. Green, *The Gospel of Luke*, New International Commentary on the New Testament, Eerdmans, Grand Rapids, 1997

John E. Hartley, *Leviticus*, Word Biblical Commentary, Volume 4, Dallas, 1992

T. R. Hobbs, *2 Kings*, Word Biblical Commentary, Volume 13, Waco, 1985

John Houghton, *The Search for God: Can Science Help?*, Lion Publishing, Oxford, 1995

Gerald L. Keown et al, *Jeremiah 26-52*, Word Biblical Commentary, Volume 27, Dallas, 1995

Hans- Joachim Kraus, *Psalms 1-59*, Fortress Press, Minneapolis, 1993

William L. Lane, *The Gospel of Mark*, Eerdmans, Grand Rapids, 1974

Tod Linafelt, *Berit Olam: Studies in Hebrew Narrative and Poetry – Ruth and Esther*, Liturgical Press, Collegeville, 1999

Andrew T. Lincoln, *Ephesians*, Word Biblical Commentary, Volume 42, Dallas, 1990

I. Howard Marshall, *The Gospel of Luke*, New International Greek Testament Commentary, Paternoster, Exeter, 1978

James L. Mays, *Hosea*, SCM Press, London, 1969

P. Kyle McCarter, Jr., *I Samuel*, The Anchor Bible, Doubleday, New York, 1980

Alister E. McGrath, *A Scientific Theology: Volume 1, Nature*, T&T Clark, Edinburgh, 2001

William McKane, *Proverbs*, SCM Press, London, 1977

Leon Morris, *The First and Second Epistles to the Thessalonians*, Eerdmans, Grand Rapids, 1991

Robert H. Mounce, *The Book of Revelation*, Eerdmans, Grand Rapids, 1977

Roland Murphy, *Ecclesiastes*, Word Biblical Commentary, Volume 23A, Word Books, Dallas, 1992

Roland E. Murphy, *Proverbs*, Word Biblical Commentary, Volume 22, Thomas Nelson, Nashville, 1998

Richard Nelson, *First and Second Kings*, John Knox Press, Louisville, 1987

H.R. Page, *The Myth of Cosmic Rebellion*, Brill, Leiden, 1996

John Polkinghorne, *Theology in the Context of Science*, SPCK, 2008.

E. C. Rust, *Nature and Man in Biblical Thought*, Lutterworth Press, London, 1953

J. N. Sanders and B. A. Mastin (eds.), *A Commentary on the Gospel According to St John*, Black, London, 1968

E. A. Speiser, *Genesis*, The Anchor Bible, Doubleday, New York, 1964

Douglas Stuart, *Hosea-Jonah*, Word Biblical Commentary, Volume 31, Thomas Nelson, Nashville, 1987

Richard Swinburne, *The Concept of Miracle*, Macmillan, London, 1970

Marvin E. Tate, *Psalms 51-100,* Word Biblical Commentary, Nelson, 1990

Charles Taylor, *A Secular Age*, Belknap/Harvard University Press, 2007

Keith Ward, *Divine Action: Examining God's Role in an Open and Emergent Universe*, Templeton Foundation Press, Philadelphia, 2007 (1990)

Keith Ward, *The Big Questions in Science and Religion*, Templeton, West Conshohocken, Pa., 2008

Gordon J. Wenham, *Genesis 1-15,* Word Biblical Commentary, Volume 1, Word Books, Waco, 1987

Claus Westermann, *Lamentations*, T&T Clark, Edinburgh, 1994

H. G. M. Williamson, *Ezra, Nehemiah*, Word Biblical Commentary, Volume 16, Waco, 1985

Ben Witherington III, *Revelation,* New Cambridge Bible Commentary, Cambridge University Press, 2003

Endnotes

1) Francis Collins, *The Language of God*, Pocket Books, Simon and Schuster UK Ltd, London, 2007.
2) Earlier scientific discoveries had been made in some of the civilisations of the East, particularly in China. This study focuses on the situation in Europe and the biblical lands.
3) Keith Ward, *Divine Action: Examining God's Role in an Open and Emergent Universe*, Templeton Foundation Press, Philadelphia, 2007 (1990), p. 74.
4) For a more detailed treatment of the slippery nature of the word "nature", see Alister E. McGrath, *A Scientific Theology: Volume 1, Nature*, T&T Clark, Edinburgh, 2001, especially chapter 3.
5) See for example, Richard Swinburne, *The Concept of Miracle*, Macmillan, London, 1970, and Keith Ward, *The Big Questions in Science and Religion*, Templeton, West Conshohocken, Pa., 2008, ch. 4.
6) For example by Marvin E. Tate, *Psalms 51-100*, Word Biblical Commentary, Nelson, 1990, p. 376.
7) W. Brueggemann, *The Message of the Psalms*, Augsburg Old Testament Studies, Minneapolis, 1984, quoted by Tate, *Psalms 51-100*, p. 381.
8) Mitchell Dahood, *Psalms III (101-150)*, Anchor Bible, Doubleday, New York, 1970, p. 265.
9) E. A. Speiser, *Genesis*, The Anchor Bible, Doubleday, New York, 1964, p. 45.
10) For example, Psalm 48.1-2 (possibly adopted from a Canaanite source – it refers confusingly to Mount Zion in the north); Isaiah 2.2-3; 56.7; 57.13; 65.11; Ezekiel 17.22-23; 20.40; Micah 4.1. The passages at Isaiah 2 and Micah 4 appear to be describing a visionary future rather than the present height of Mount Zion.
11) Hans- Joachim Kraus, *Psalms 1-59*, Fortress Press,

Minneapolis, 1993, p. 222.

12) Joseph Blenkinsopp, *Isaiah 1-39,* The Anchor Bible, Doubleday, New York, 2000, p. 438.
13) Psalms 14 and 53 are close parallels, their main difference being in the name used for God. Psalm 14 mostly uses Yahweh, "the Lord", while Psalm 53 uses Elohim, "God".
14) Dahood, *Psalms III (101-150),* p. 276.
15) On the Lord laughing at his enemies see also Psalm 37.13 and 59.8.
16) Richard Nelson, *First and Second Kings,* John Knox Press, Louisville, 1987, p. 141.
17) Kraus, *Psalms 1-59,* p. 203.
18) See John Day, *Yahweh and the Gods and Goddesses of Canaan,* Sheffield Academic Press, 2000, and H.R. Page, *The Myth of Cosmic Rebellion,* Brill, Leiden, 1996.
19) Joseph Blenkinsopp, *Isaiah 56-66,* The Anchor Bible, Doubleday, New York, 2003, pp. 42-44, 294.
20) *Ibid,* p. 262.
21) *Ibid,* p. 295.
22) The descent of God to view the tower of Babel in Genesis 11.5 is described by Gordon J. Wenham as "no more proof of the author's primitive anthropomorphic view of God than is God asking Adam and Eve where they were hiding in the garden an indication of his ignorance." (*Genesis 1-15,* Word Biblical Commentary, Volume 1, Word Books, Waco, 1987, p. 240). The cumulative weight of spatial references to God's location suggests, however, that something more than anthropomorphism is involved.
23) Nelson, *First and Second Kings,* p. 53.
24) Dahood, *Psalms III (101-150),* p. 289.
25) William McKane, *Proverbs,* SCM Press, London, 1977, pp. 479-480.
26) Roland E. Murphy, *Proverbs,* Word Biblical Commentary, Volume 22, Thomas Nelson, Nashville, 1998, p. 114.

27) Joel B. Green, *The Gospel of Luke,* New International Commentary on the New Testament, Eerdmans, Grand Rapids, 1997, p. 861. It seems unnecessary to *contrast* this theological interpretation with a view that 'upward' relates to a cosmological model, as Green does in footnote 12 on the same page.
28) I. Howard Marshall, *The Gospel of Luke,* New International Greek Testament Commentary, Paternoster, Exeter, 1978, p. 909.
29) James D. G. Dunn, *The Acts of the Apostles,* Epworth Press, Peterborough, 1996, p.13.
30) Gordon Fee argues that the description of Christ as "of heaven" does *not* convey the meaning that Christ *came* from heaven – rather the sense is that Christ is of "heavenly" nature compared with Adam's (and all humans') earthly nature. The word describes character rather than origins. Gordon D. Fee, *The First Epistle to the Corinthians,* New International Commentary on the New Testament, Eerdmans, Grand Rapids, 1987, pp. 792-793.
31) Andrew T. Lincoln, *Ephesians,* Word Biblical Commentary, Volume 42, Dallas, 1990, pp. 20, 62.
32) G. K. Beale, *The Book of Revelation,* New International Greek Commentary, Eerdmans, Grand Rapids, 1999, p. 523.
33) Robert H. Mounce, *The Book of Revelation,* Eerdmans, Grand Rapids, 1977, p. 378.
34) Ben Witherington III, *Revelation,* New Cambridge Bible Commentary, Cambridge University Press, 2003, p. 254.
35) Albright and Mann give the lively translation "fast approaching" for the more conventional "at hand". W. F. Albright and C. S. Mann, *Matthew,* The Anchor Bible, Doubleday, New York, 1971, pp. 24-25, 38-39.
36) The quotation "we too are his offspring" is from Aratus' *Phaenomena*; other sources have been proposed for the other quotations in Paul's speech, but there is no agreement on

their provenance.

37) See, for example, Oscar Cullmann, *Christ and Time*, SCM Press, London, 1962.
38) E. C. Rust, *Nature and Man in Biblical Thought*, Lutterworth Press, London, 1953.
39) *Ibid*, p. 17.
40) *Ibid*, p. 18.
41) *Ibid*, pp. 69-70.
42) *Ibid*, p. 259.
43) Noah Efron, "Early Judaism", in John Hedley Brooke and Ronald L. Numbers (eds.), *Science and Religion around the World*, Oxford University Press, 2011, ch. 1.
44) See also the permanence of the earth at Psalm 96.10; 104.5.
45) C. E. B. Cranfield, *Romans: A Shorter Commentary*, T&T Clark, Edinburgh, 2001 (1985), p. 196.
46) See Genesis 7.11, 8.2; Psalm 104.3; 148.4.
47) Leslie C. Allen, *The Books of Joel, Obadiah, Jonah and Micah*, Eerdmans, Grand Rapids, 1976, p. 224.
48) Deuteronomy 32.10-13; Psalm 17.8; 36.7; 57.1; 61.4; 63.7; Ruth 2.12.
49) Isaiah 5.17; 13.19-22; 14.22-23; 17.2; 27.10; 32.14; 34.11-15. In the later section of the book of Isaiah, some of these same wild animals, including jackals and ostriches, will "honour" God when he renews the land and brings the people back from captivity (Isaiah 43.20) – presumably these animals will abandon the ruins and return to their former places in the real wilderness.
50) Fee, *Corinthians*, p. 406, n. 52.
51) *Ibid*, p. 408.
52) Nelson, *First and Second Kings*, p. 56; Simon J. DeVries, *1 Kings*, Word Biblical Commentary, Volume 12, Waco, 1985, p. 121.
53) For further references to hail as God's judgment see also Exodus 9.22-26; Job 38.22-23; Isaiah 28.2, 17; 30.30; 32.19

(NRSV footnote); Ezekiel 13.13 and Haggai 2.17.

54) Also 1 Samuel 7.10; 12.17-18; 2 Samuel 22.14.

55) R. T. France, *The Gospel of Matthew*, Eerdmans, Grand Rapids, 2007, pp. 334-337.

56) Michael V. Fox, *Proverbs 1-9*, The Anchor Bible, Doubleday, New York, 2000, p.159.

57) The attribution of God's action to his "wisdom" is also found in Psalm 104.24, and the role in creation of God's power, wisdom and understanding is also found in Jeremiah 10.12 and 51.15.

58) Modern theories of climate change suggest that human activities such as the burning of fossil fuels can indeed affect the environment, including phenomena such as rainfall or the frequency of storms. But this is a long way from the biblical view of God's control of rainfall and drought. The modern view of climate change is that it is a much longer, slower process than the divine actions described in the Bible. Moreover, climate change potentially affects the whole world, not solely a small territory or nation which may have been misbehaving.

59) This hint in Jeremiah 14.22 of a perceived operation in nature independent of God (which Jeremiah himself rejects) seems to be repeated in another such hint in Mark 4.28, where Jesus says that seed grows "of itself" – the Greek word is αυτοματη, "automatically". Perhaps the meaning here is "without human intervention", but nevertheless the word raises intriguing possibilities about the perceived role of God in plant growth.

60) John Houghton, *The Search for God: Can Science Help?*, Lion Publishing, Oxford, 1995, p. 154.

61) *Ibid*, p. 167.

62) Is there a hint of humour or mystery in Genesis 3.16? God tells the woman that she will bear children in pain, and that her husband will rule over her, and "yet your desire shall be

for your husband". Is the (male?) author expressing some surprise that women should accept this combination of conditions?

63) For the priority of males in the worship of Yahweh, see also Exodus 23.17; 34.23; Deuteronomy 16.16.

64) John I. Durham, *Exodus*, Word Biblical Commentary, Volume 3, Waco, 1987, p. 265.

65) In Job 42.15 an alternative practice is described: following Job's restoration to prosperity, his daughters receive an inheritance along with his sons. For a more detailed discussion of variations in inheritance and landholding, see Timothy R. Ashley, *The Book of Numbers*, Eerdmans, Grand Rapids, 1993, pp. 540-546.

66) Philip J. Budd, *Numbers*, Word Biblical Commentary, Volume 5, Waco, 1984, p. 300.

67) *Ibid*, p. 324.

68) H. G. M. Williamson, *Ezra, Nehemiah*, Word Biblical Commentary, Volume 16, Waco, 1985, p. 157.

69) The language in 1 Corinthians 11 is in terms of the "head". This may refer to authority, but most commentators prefer the sense of the head as a "source", hence a reference to man as the source of the woman in the creation story in Genesis 2. But are commentators attempting to soften the distinction between modern and ancient views of gender?

70) Tod Linafelt in *Berit Olam: Studies in Hebrew Narrative and Poetry – Ruth and Esther*, Liturgical Press, Collegeville, 1999, p. 77. Linafelt goes on to castigate the neglect of any discussion, either in the biblical text or in subsequent interpretation, of Ruth's desires in the matter.

71) Psalm 71.6; 113.9; 115.14; 128.1-4; Isaiah 66.9; Jeremiah 1.5; Ezekiel 36.37.

72) Psalm 111.5 also attributes to God the provision of food.

73) P. Kyle McCarter, Jr., *I Samuel*, The Anchor Bible, Doubleday, New York, 1980, p. 84.

74) Possible textual variants suggest that this judgment may have been because the ark had been handled by people who were not priests, see McCarter, *I Samuel*, p. 131.
75) In 1 Chronicles 21.1 it is Satan who incites David to launch the census, evidence of a process to distance God from the sources of evil.
76) T. R. Hobbs, *2 Kings*, Word Biblical Commentary, Volume 13, Waco, 1985, p. 237.
77) Nelson, *First and Second Kings*, p. 264.
78) Similar accusations that God was the direct cause of affliction are found at Exodus 5.22, Numbers 11.11 and 1 Kings 17.20. DeVries suggests that 1 Kings 17.20, in the story of Elijah and the widow's son, is a secondary addition as it places in Elijah's mouth "the woman's bad theology to the effect that Yahweh might actually kill little boys." (DeVries, *1 Kings*, p. 222). Nelson, however, writes of Elijah as "one deeply anguished by divine ambiguity" (*First and Second Kings*, p. 112). Whether the verse is secondary or not, it remains the case that some in the period, whether desperate mothers, anguished prophets or mere editors, held to this "bad theology".
79) Blenkinsopp comments: "the traditional belief that Yahweh is the source of everything that happens, whether good or ill, was still widely, though not universally, accepted as unproblematic (see, for example, Amos 3.6; 5.18-20; Isaiah 41.23)." Joseph Blenkinsopp, *Isaiah 40-55*, The Anchor Bible, Doubleday, New York, 2002, p. 250.
80) Duane Garrett and Paul R. House, *Song of Songs/ Lamentations*, Word Biblical Commentary, Volume 23B, Thomas Nelson, Nashville, 2004, p. 419.
81) Some textual critics here prefer "speak of him" rather than "speak against him".
82) Matthew 18.8; 25.41
83) Matthew 13.42; 22.13; 24.51; 25.30; Luke 13.28

84) Matthew 13.30; John 15.6
85) J. N. Sanders and B. A. Mastin (eds.), *A Commentary on the Gospel According to St John*, Black, London, 1968, p. 237.
86) Alternatively, the past tense may be used to describe the certainty of future judgement. Leon Morris, *The First and Second Epistles to the Thessalonians*, Eerdmans, Grand Rapids, 1991, p. 85.
87) Douglas Stuart, *Hosea-Jonah*, Word Biblical Commentary, Volume 31, Thomas Nelson, Nashville, 1987, p. 105.
88) Also Jeremiah 10.18; 13.24; 16.13; 27.10, 15.
89) Also Jeremiah 15.2, 9; 24.10; 27.8; 29.17-18; 34.17.
90) Gerald L. Keown et al, *Jeremiah 26-52*, Word Biblical Commentary, Volume 27, Dallas, 1995, pp. 235-236.
91) Claus Westermann, *Lamentations*, T&T Clark, Edinburgh, 1994, pp. 149-150.
92) For further extended treatment in the book of Ezekiel of God's direct action in Israel's disasters see Ezekiel 6.3-5; 7.3-9, 21-24; 11.8-10; 12.13-15; 14.12-23; 15.6-8; 20.47-48; 21.1-17.
93) Kraus, *Psalms 1-59*, p. 542.
94) Mark 8.31-32; 9.9-10, 31-32; 10.32-34; cf. Matthew 16.21-22; 17.12, 22-23; 20.18-19, 22, 28; 26.2, 12, 21-24; Luke 9.22, 31, 44; 12.50; 18.31-33.
95) Mark 14.29-31, 72; cf. Matthew 26.34; John 13.38.
96) Acts 11.27-30; 20.23; 21.9-14; 1 Corinthians 12.10; 14.1-5; Ephesians 4.11; 1 Timothy 1.18; 4.14; Revelation 1.1-3. In some of these texts prophesy refers to exhortation and encouragement, not to foretelling the future.
97) Blenkinsopp, *Isaiah 1-39*, pp. 412, 427.
98) Brevard S. Childs, *Isaiah*, Westminster John Knox Press, Louisville, 2001, pp. 231-232.
99) Albright and Mann point out that the text does not state that Peter actually went and caught such a fish and suggest that the saying about the coin in the fish's mouth may have originated as a parable. Albright and Mann, *Matthew*, pp. cxxx,

213.

100) *Ibid*, p. 182.

101) See Leviticus 19.26; Deuteronomy 18.10; 1 Samuel 15.23; 28.7-25; 2 Kings 17.17; 21.6; 23.24; 2 Chronicles 33.6; Isaiah 19.3. Hobbs makes the point that while our modern society is generally suspicious of divination because it is "spooky" or "abnormal", for the Hebrews the motivation was a fear of religious apostasy (*2 Kings*, p. 235).

102) Judges 1.1-2; 20.18, 23, 26-28 (the presence of the ark of the covenant was significant in this case, v27); 1 Samuel 10.22 (not a yes/no answer in this case); 14.37-41; 23.2-4, 11-12; 28.6, 15 (refer to God answering by prophets or by dreams as well as by Urim); 30.8; 2 Samuel 2.1; 5.19, 23; 21.1 (again, a more detailed answer than yes or no). Extended examples of seeking the will of God through prophets are found at 1 Kings 22.5-28, 2 Kings 22.13-20, 2 Chronicles 18.4-27 and 34.18-28, while 1 Chronicles 21.30 and 2 Chronicles 1.5 suggest that the place to obtain God's answers was before the altar of the Lord.

103) Murphy, *Proverbs*, p. 137.

104) McKane, *Proverbs*, p. 499.

105) The passage is repeated at Jeremiah 51.15-19.

106) The same thought is also expressed in Jeremiah 33.20-26.

107) Luke's version is somewhat less universal: "What is impossible for mortals is possible for God."

108) The incongruous postscript to the Book of Job, in which not only his wealth but also his seven sons and three daughters are restored to him, undermines the whole tragic theme of the rest of the book, namely that there is *no* correlation between righteousness and prosperity.

109) Another less frequently noted example of Jesus being angry occurs at Mark 3.5, when he was confronted by the unwillingness of those in the synagogue to see a man healed on the Sabbath.

110) A good source of information on biblical references to natural phenomena, based on personal knowledge of the biblical region, is Yehuda Feliks, *Nature and Man in the Bible*, Soncino Press, London, 1981.

111) McKane, *Proverbs*, pp. 653-655.

112) James L. Mays, *Hosea*, SCM Press, London, 1969, pp. 9-10.

113) See discussion in Richard J. Bauckham, *Jude, 2 Peter*, Word Biblical Commentary, Volume 50, Waco, 1983, p. 89.

114) John E. Hartley, *Leviticus*, Word Biblical Commentary, Volume 4, Dallas, 1992, p. 162.

115) Duane L. Christensen, *Deuteronomy 21.10 – 34.12*, Word Biblical Commentary, Volume 6B, Thomas Nelson, Nashville, 2002, p. 508.

116) In connection with this verse it is sometimes suggested that the north wind does not in fact bring rain in Israel. Winter rain comes from the north-west, in contrast to summer dry heat from the south-east, so the use of north in Proverbs 25.23 may be a simple approximation.

117) Roland Murphy, *Ecclesiastes*, Word Biblical Commentary, Volume 23A, Word Books, Dallas, 1992, p. 109.

118) Judges 13.6 (Manoah's wife); Isaiah 6.2 (seraphs, the only biblical occurrence of such beings).

119) For reference to the appearance of groups of angels see Genesis 18-19; 32.1; Deuteronomy 33.2-3; Psalm 78.49; 91.11-12; 103.20; 148.2; Luke 2.13-15.

120) For example Genesis 16.7-13 (Hagar); 18-19 (Abraham); Judges 6.11-16 (Gideon); 2 Samuel 24.17 (David).

121) Genesis 32.30; Exodus 33.20; Deuteronomy 4.33; 5.24, 26; Judges 6.22-23; 13.22.

122) Ashley, *The Book of Numbers*, p. 225.

123) For example by Charles Taylor, *A Secular Age*, Belknap/Harvard University Press, 2007.

124) Some examples will show that these problems are relatively trivial but nevertheless present:

Was Noah commanded to take into the ark one pair of animals (Genesis 6.19-20; 7.8) or seven pairs (Genesis 7.2-3)?

Was Goliath killed by David (1Samuel 17) or by Elhanan son of Jaareoregim the Bethlehemite (2Samuel 21.19)? [The Book of Chronicles, which was written later than 1 and 2 Samuel, tidies this up by saying that Elhanan "killed Lahmi the brother of Goliath" (1 Chronicles 20.5), but there is a mistake here somewhere.]

The census under Quirinius, used by Luke 2.2 to locate Jesus' birth in Bethlehem, actually took place in 6 AD.

Did Jesus meet Bartimaeus as Jesus was leaving Jericho (Mark 10.46) or as he was approaching the city (Luke 18.35)?

Matthew 23.35 confuses Zechariah son of Berachiah (the prophet) with Zechariah son of Jehoiada (the real murder victim).

Was the resurrection witnessed by one woman (Mark 16.9, John 20.1-18) or two (Matthew 28.1-10) or three (Mark 16.1-8) or more (Luke 23.55 - 24.10)?

When Paul was converted on the way to Damascus, did his companions hear the voice (Acts 9.7) or not (Acts 22.9)?

More generally, the gospel accounts of the teachings of Jesus, while pointing back to the same original source, are sufficiently varied to warn us that, even allowing for thei translation into Greek, we are not reading an accurat version of his original words.

125) John Polkinghorne, *Theology in the Context of Science*, SPC 2008, p. 86.

126) In reality, the Big Bang explanation for the origin of universe does not extend as far back as the abso beginning (t=0). The unexplained interval is minut human time scales, but remains as a lacuna betwee scientific and religious accounts of the origin o universe. The interval also leaves space for further sta the scientific explanation.